Improving farm animal welfare

Improving farm animal welfare

Science and society working together: the Welfare Quality approach

edited by:

Harry Blokhuis

Mara Miele

Isabelle Veissier

Bryan Jones

Wageningen Academic
P u b l i s h e r s

ISBN: 978-90-8686-216-0
eISBN: 978-90-8686-770-7
DOI: 10.3920/978-90-8686-770-7

Cover illustration: Raymond Nowak

First published, 2013

© **Wageningen Academic Publishers**
The Netherlands, 2013

This work is subject to copyright. All rights
are reserved, whether the whole or part of
the material is concerned. Nothing from this
publication may be translated, reproduced,
stored in a computerised system or published
in any form or in any manner, including
electronic, mechanical, reprographic
or photographic, without prior written
permission from the publisher:
Wageningen Academic Publishers
P.O. Box 220
6700 AE Wageningen
The Netherlands
www.WageningenAcademic.com
copyright@WageningenAcademic.com

Table of contents

Acknowledgements

This book could not have been written without the commitment and efforts of all the scientists, members of the public, farmers, retailers, NGO's and other stakeholders who contributed to the success of the Welfare Quality® project.

The editors would like to specifically acknowledge all the contributors to this book and last but not least the critical role of the European Union and the 6th Framework Programme (contract No. FOOD-CT-2004-506508). This important step forward in animal welfare science would not have been possible without their vision and support. The text represents the authors' views and does not necessarily represent a position of the Commission who will not be liable for the use made of such information.

Harry Blokhuis
Bryan Jones
Isabelle Veissier
Mara Miele

Chapter 1. Introduction

Harry Blokhuis, Bryan Jones, Isabelle Veissier and Mara Miele

In this book we address the complex and often controversial issues surrounding the assessment and improvement of farm animal welfare from production to consumption by addressing a crucial question, i.e. what is a good quality of life for a farmed animal? Based on the approach taken in a large, multi-disciplinary EU funded research project called Welfare Quality®, the book discusses the pressing need for reliable and holistic science based welfare assessments and the importance of establishing a fruitful dialogue between science and society. It then crucially describes the establishment of the Welfare Quality® Principles and Criteria for good welfare, the development of workable welfare assessment and scoring systems for cattle, pigs and chickens as well as practical ways of improving selected aspects of the animals' quality of life. In short, this book synthesises the huge body of work carried out by the largest ever international network of scientists and stakeholders in Welfare Quality® and describes why particular paths were chosen, some of the obstacles encountered and how they were overcome, as well as selected outputs and major achievements. It also clearly sets out what still needs to be done and presents selected strategies and technologies (automation, proxy indicators, targeting of risk factors, etc.) designed to ensure the continued improvement of welfare and its assessment.

It is almost 10 years since the first aims and approaches of the Welfare Quality® project were formulated by a small group of committed scientists in response to the European Commission's call for proposals with the ultimate objective 'of improving animal production methods that take into account consumer demands for high standards of animal welfare, health and food quality'. New knowledge should be generated regarding objective indicators of welfare status and amelioration of welfare problems. The project should link together a wide range of stakeholders and stimulate a science-society dialogue on welfare issues in farming. To address this call we focussed on a multidisciplinary approach and the integration of European strength in the field of animal welfare.

Our research proposal was successful and Welfare Quality® was financed by the Commission under the European 6th Framework Programme for Research and Technological Development (FP6). The full title of the project was 'Integration of animal welfare in the food quality chain: from public concern to improved welfare and transparent quality'. The project itself began in May 2004 and ended in December 2009. By then it had become the largest piece of integrated research work ever carried out on animal welfare in Europe. The partnership involved approximately 200 scientists representing 43 institutes and universities in 13 European and 4 Latin American countries (see list of partners in Appendix 1). Collectively, the participants offered a broad range of specialist expertise in several disciplines ranging from the social sciences to numerous branches of biology. Indeed, the Welfare Quality® partnership included many of the leading European experts in farm animal welfare.

A strong management structure was established as early as the proposal writing stage and this helped ensure that the project hit the ground running. A Steering Committee (SC) took ultimate responsibility for the overall management of the project and was supported by a Management Team consisting of the SC and the leaders of each of the four main sub-projects. These two bodies were assisted by a professional project office which developed, in close co-operation with the SC, dedicated and effective management and administration tools for the day to day running of the project. Two advisory bodies were also established to work with the SC and other project participants. The Advisory Committee consisted of an ethicist and representatives from farmer, animal breeding, retail, food service, certification and veterinary organisations. They provided advice on the relevance, timeliness and progress of the work and incoming proposals and on the inclusion of specific issues and strategies. A Scientific Board consisted of international experts whose collective knowledge covered all aspects of the project with a remit to help assess the scientific and technical quality of the work.

Within FP6, the Thematic Priority 'Food Quality and Safety' was defined with the primary objective to improve the health and well-being of European citizens through ensuring a higher quality of food. It was becoming increasingly recognised at this time that consumers expected their animal-related products, especially food, to be produced and processed with greater respect for the welfare of the animals (Blokhuis et al., 2003). Thus, their perception of food quality was clearly determined not only by the overall nature and safety of the end product but also by the welfare status of the animals from which the food was produced. In other words, the welfare of farm animals became more and more a part of an overall concept of 'food quality'. Since consumers are the ones who buy the product it was clear that their demands should be the major drivers for change in the whole production chain ('from fork to farm').

The public was concerned not only about the status of animal welfare but also about the relative lack of clear and easily understood information that would allow them to make informed choices about the animal products they buy (Blokhuis et al., 2003). Other very influential drivers of efforts designed to improve welfare included government and industry. Firstly for example, the fact that poor welfare can result in poor animal health was widely accepted and this led policy makers and risk managers to adopt a new approach by integrating animal welfare, animal health and food safety

(Blokhuis *et al.*, 1998, 2008). Secondly, it had also become increasingly recognised that productivity, product quality and profitability are often reduced if the animals' welfare is compromised (Hemsworth and Coleman, 1998; Jones, 1997) thereby pointing to the strong economic as well as ethical and safety reasons for assessing, safeguarding and improving farm animal welfare.

An integrated programme of research was clearly required to help satisfy a number of needs including: to improve animal husbandry and animal welfare from housing to slaughter, to assure European citizens and other stakeholders of the quality of the food products, to provide the related information they demanded regarding the welfare status of farm animals, and to safeguard the sustainability of European agriculture. The Welfare Quality® project addressed this growing societal need of consumers and citizens for a high welfare quality and increased transparency of production. Our collaborative efforts were designed to ensure the clear integration of animal welfare in the food quality chain and involved animal and social scientists, mathematicians, farmers, processors, slaughter house managers, retailers, NGOs, members of the public, etc. In short, the Welfare Quality® project provided an excellent example of science and society working together to improve farm animal welfare.

A reliable system for assessing animal welfare on farm and at slaughterhouses was identified as one of the primary requirements (Blokhuis *et al.*, 2003). Although animal welfare can be a difficult concept to define there was general agreement within the scientific community about the broad terms of what represents good animal welfare. This consensus had been synthesised and elegantly expressed by the UK's Farm Animal Welfare Council in the 'Five Freedoms' (FAWC, 1992). The Welfare Quality® consortium recognised that to be widely accepted a balanced welfare assessment system has to satisfy public, industry, political and scientific concerns. Therefore, Welfare Quality® researchers expanded these earlier approaches to animal welfare definition and assessment and established a holistic concept covering the different domains of animal welfare. In a critical first step, the views of consumers, industry, farmers, legislators and scientists were drawn together to establish four principles which were considered essential to safeguard and improve farm animal welfare: good housing, good feeding, good health and appropriate behaviour. Twelve distinct but complementary criteria for good welfare were then defined within these four principles. These principles and criteria complemented and extended the Five Freedoms and they provided the solid platform that was needed to build the Welfare Quality® assessment system.

Hitherto most of the protocols used by various schemes to monitor animal welfare on farms and at slaughter relied heavily on design- or resource-based measures. In contrast, and as early as the proposal writing stage, the Welfare Quality® team made a deliberate decision to focus on the state of the animal rather than just the nature and quality of its living conditions, although of course these have a large impact on the actual welfare status of the animal (Blokhuis *et al.*, 2003). Therefore, Welfare Quality® focussed primarily on the development of animal-based measures, which attempt to assess welfare from the animals' point of view. This viewpoint formed the basis for the development of an assessment tool that we believe can play a central, on-going and evolving role in many processes designed to improve farm animal welfare as well as in regulatory efforts in this area.

Reporting the detailed results of the assessment measures to the farmer is a central part of the Welfare Quality® vision. The subsequent provision of scientifically sound knowledge-based advice on appropriate remedial measures for specific welfare problems is an integral component of this feedback process. In this context Welfare Quality® researchers identified a number of practical ways of improving the welfare of pigs, cattle and poultry. Welfare Quality® scientists are continuing to build a Technical Information Resource (TIR) on welfare improvement strategies that have been developed both within and outside the Welfare Quality® project (Chapter 8; Jones and Manteca, 2009).

Apart from its considerable output, the many multi-national and multi-disciplinary efforts engendered in the Welfare Quality® project made an immense contribution to the stimulation of a widespread integration of research teams in Europe and beyond. Examples of the various disciplines involved include biology, biochemistry, ethology, psychology, physiology, animal science, animal husbandry, mathematics, ethics, economics and social sciences. Coordinated and collaborative efforts are not only required by the transnational perspective of the animal welfare issues but are also considered essential in order to overcome the fragmentation of research in Europe along national and institutional barriers. The Welfare Quality® project clearly addressed these issues which are central to the concept of the European Research Area (http://ec.europa.eu/research/era/index_en.htm).

There is little point in producing a large amount of potentially valuable information and then leaving it to sit on a shelf. Therefore, the Welfare Quality® team devoted substantial effort to the communication and dissemination of its findings and recommendations to a broad audience. The disseminated material included: a dedicated website; welfare assessment protocols for each of the seven animal groups; three Welfare Quality® stakeholder conferences and their related proceedings; a series of twelve Welfare Quality® reports on various aspects of the project and its findings; several easily understandable fact sheets covering some of the innovative strategies, knowledge and recommendations concerning farm animal welfare that were engendered by the project; a Technical Information Resource for practical welfare improvement strategies; many popular publications and scientific papers; numerous media interviews and newspaper articles, and the Welfare Quality® DVD (see www. welfarequality.net and www.welfarequalitynetwork.net).

Although the Welfare Quality® project finished at the end of 2009 its legacy continued in the form of two related projects: the European Animal Welfare Platform (EAWP) and the Welfare Quality Network (WQN) (Chapter 10) as well as numerous smaller projects that utilise or further develop Welfare Quality® results. A new initiative – the Coordinated European Animal Welfare Network (EUWelNet) also started near the end of 2012 (Chapter 10).

Rather than serving as an all-encompassing report this book aims to provide a brief but comprehensive overview of the thinking behind the work as well as the efforts of all the scientists, farmers, retailers and other stakeholders who contributed to the success of this very large Welfare Quality® project. It is also intended to ensure that the many important results and recommendations generated in the project are gathered together in one easily accessible source. An interested reader can easily find greater detail and in-depth discussion by following the relevant links and numerous references cited in the following nine chapters.

References

Blokhuis, H.J., Hopster, H., Geverink, N.A., Korte, S.M. and Van Reenen, C.G. (1998). Studies of stress in farm animals. Comparative Haematology International, 8, 94-101.

Blokhuis, H.J., Jones, R.B., Geers, R., Miele, M. and Veissier, I. (2003). Measuring and monitoring animal welfare: Transparency in the food product quality chain. Animal Welfare, 12, 445-455.

Blokhuis, H.J., Keeling, L.J., Gavinelli, A. and Serratosa, J. (2008). Animal welfare's impact on the food chain. Trends in Food Science and Technology, 19, 75-83.

FAWC (1992). Farm animal welfare council: FAWC updates the five freedoms. Veterinary Record, 131, 357.

Hemsworth, P.H. and Coleman, G.J. (1998). Human-livestock interactions: the stockperson and the productivity of intensively farmed animals. CAB International, Wallingford, United Kingdom. 152 pp.

Jones, R.B. (1997). Fear and distress. In: Appleby M.C. and Hughes B.O (eds.) Animal Welfare. CAB International, Wallingford, UK, pp. 75-87.

Jones, R.B. and Manteca, X. (2009). Best of breed. Public Science Review, 18, 562-563.

Chapter 2. Changes in farming and in stakeholder concern for animal welfare

Mara Miele, Harry Blokhuis, Richard Bennett and Bettina Bock

2.1 Introduction

The present chapter aims to describe and discuss the socio-economic developments and the related scientific advancements that formed the background and context of the Welfare Quality project. This project was clearly a 'child of its time'. In the last decades animal welfare issues have attracted growing public attention and research in this area has grown to become a mature scientific discipline that is capable of addressing the new societal, political and market demands for more animal friendly types of production. This chapter specifically addresses the changes in animal farming that have occurred in the last fifty years as well as the emergence of animal welfare science. It then examines EU policy and legislation on the protection of farm animal welfare and changes in farmers' attitudes and stakeholders' concerns regarding animal welfare in Europe. The final section is dedicated to the analysis of the rise of public concern for farm animal welfare as well as the conditions for the emergence of consumer demand for animal friendly products.

2.2 Changes in animal farming in the post World War II era and the emergence and role of animal welfare science

Livestock production is the world's largest user of land and accounts for almost 40% of the total value of agricultural production. In industrial countries this figure is more than 50% and in developing countries it is rapidly rising from the current 33% (Bruinsma, 2003). Livestock production has achieved this prominence in recent times, with the strong specialisation of animal husbandry for food production predominantly in industrial countries, while in developing countries farmed animals are still used for work (e.g. transport, haulage, ploughing) and other purposes[1] as well as for food. Global meat production has tripled from 47 million tons in 1980 to 139 million tons in 2002 (Steinfeld *et al.*, 2006) and it is expected to double by 2030 (Bruinsma, 2003) due to the rising demand especially in developing countries where people are adopting western diets and styles of consumption (Hendrickson and Miele, 2009).

---

[1] As 'piggy bank'/saving, and for farm labour- ploughing especially.

Mara Miele, Harry Blokhuis, Richard Bennett and Bettina Bock

Western Europe (with North America and East Asia) are the regions with the highest industrialisation of animal production, the highest concentration of animals reared for food in the world, and with the highest levels of output per animal unit (Ruttan, 1998). These outcomes reflect the major changes that have taken place in the (Western) European animal production sector during the latter part of the 20th century and, as Hendrickson and Miele (2009) have argued, while these changes contributed to an increased food security in Europe they also affected the welfare of farmed animals and raised public concerns about farming (Blokhuis *et al.*, 1998; Fraser, 2008). The changes in farm structural and enterprise characteristics were heavily influenced by the European Union´s Common Agricultural Policy (Winter *et al.*, 1998) that promoted mechanisation and specialisation of farming. Whereas the overall number of farms fell markedly there was a significant rise in the number of animals per farm (Porcher, 2006). For instance, the average number of laying hens in the Netherlands increased from about 17.7 million in 1976 to 30 million in 1997 even though the number of farms with laying hens decreased over that period from about 13,750 to 2,340 (CBS, 1998; LEI/CBS, 1984). Another example is the number of farms raising meat-producing species such as chickens, pigs and cattle in Denmark where the percentage decrease per year between 1970 and 2000 was 3.1, 3.0 and 2.6 respectively (Fraser, 2005).

Moreover, during this period the level of production per animal increased enormously. Between 1960 and 1995 the milk produced per cow in the Netherlands increased by almost 60% and, due to continuous genetic selection for increased growth rate and better food conversion in broiler chickens, the time required to reach a live weight of 1.8 kg decreased from 91 days in 1954 to only 37 days in 1996 (Vos, 1997). It is now thought that broilers reach a set '5-week slaughter weight' almost half a day sooner each year (P. Cook, Food Animal Initiative, personal communication).

Furthermore, most animals nowadays live in highly specialised farms that concentrate on a specific type of production such as milk, poultry meat or veal. Such specialisation is also apparent within a specific production chain. For instance parent stock that produce hens for egg production are kept on specialised farms, the eggs are then transported to hatcheries where laying hen chicks are hatched, these are then transported to farms that specialise in the rearing of chicks to about 17 weeks, at which time the young hens are transported to the actual egg producing units.

These changes were made possible and were also stimulated by parallel developments in housing systems and management practices as well as related mechanisation and other technological developments.

Housing conditions, especially for pigs and poultry, changed profoundly where low-density systems (often outdoor) were replaced by housing systems (often indoor) characterised by high animal density with minimal living space for the individual and a very barren environment (Blokhuis, 1999). These systems allowed a high degree of mechanisation, such as automatic manure removal, egg collection, climatic control, etc. thereby decreasing the labour requirement. The latter contributed to an enormous decrease in the workforce employed[2] in EU (12 Member States) agriculture which fell from 13.5% to 5.5% between 1970 and 1994 (Grant, 1997).

Thus animal production intensified enormously over the last 50 years or so, especially in developing countries that, since the 90s, produce more meat than developed countries (Fraser, 2008). And global meat production is projected to more than double from 229 mt in 1999-2001 to 465 mt in 2050, while milk output is set to climb from 580 to 1,043 mt (OECD-FAO, 2011).

This intensification not only enables a large increase in production volume but also increased food security/independence in Europe and other industrialised countries, with significant changes in diet (high increase in meat consumption, as shown in Table 2.1). However, the barren housing conditions, high production levels and profound mechanisation also caused growing concern and fierce societal debate regarding the welfare of the animals. The concerns for the quality of life of a rapidly growing number of animals used in food production, firstly voiced by some animal advocates and pioneer scientists but quickly grown into a new social movement, also paralleled the development of a specific area of research that eventually integrated expertise from several disciplines, including veterinary science, biology, physiology, neuroscience, ethology and ethics, and gave birth to what is now known as 'animal welfare science'.

Table 2.1. Changes in meat consumption kg per person per annum in the last 40 years (FAO, 2004.

Region	1961	2002
Europe	56	89
USA	89	124
China	4	54

[2] Even more evident in the crops sector.

2.3 Development of animal welfare science

The scientific study of animal welfare is a relatively young but well established scientific discipline (c.f. Millman *et al.*, 2004). The area has developed over the last four or five decades and continues to expand and diversify to meet new challenges and new opportunities. It is generally accepted that animal welfare is about the animal itself, and the increasing integration of fundamental biological sciences is contributing towards a greater understanding of the link between the animal's biology and its welfare state. Parallel to the basic research there is a rapidly growing area of applied animal welfare research directed towards continued improvement of ways to measure the welfare of farmed animals in practice (on farm, during transport and at slaughter) and to the development of practical strategies designed to enhance welfare.

Blokhuis *et al.* (2008) illustrated the development of animal welfare science with the dramatic increase in the total numbers of publications on animal welfare and animal wellbeing. They reported that a literature search in Web of Science in 2007 generated over 35,000 'hits', and 46% of the publications could be attributed to authors with an address in Europe, 38% were from North America, 10% from Asia, 3% from Australia and New Zealand and 3% from South America. An enormous output such as this reflects the importance of animal welfare globally and the leadership of European research.

The brief overview of animal welfare science in the context of the food chain that follows is mainly derived from the paper published by Blokhuis *et al.* in 2008[3]. That paper identified four main influential areas representing the contributions made respectively by research related to the animals themselves, their housing and husbandry, the role of societal concerns and animal welfare policy.

The origin of farm animal welfare science dates back to the 1960's (Brambell Committee, 1965; Harrison, 1964). Since that time the distinction between animal protection (*what people are allowed to do to animals*) and animal welfare (*probing the animal's own experience of its situation*) has grown and it is now accepted that animal welfare science is largely about the assessment of the animal's own experience. The two most widely quoted definitions (Broom, 1996; Duncan, 1993) state that animal welfare concerns an animal's ability to cope with its environment and, because the concept is only applied to sentient animals, animal welfare focuses on how animals feel. Thus, fundamental research in this area usually reflects the need to get '*inside the mind*' of the animal.

---

[3] This section draws on the significant contribution of Linda Keeling to this previously published paper.

Improving farm animal welfare

The dominant research disciplines in this area are the behavioural and physiological sciences. For instance, the way the body responds to stressful stimulation is a key area, with clear and important consequences for productivity, product quality and profitability (Jones, 1997; Jones and Boissy, 2011; Gregory, 1998). The work often involves studies of animals' responses to exposure to acute and chronic stressors, especially the functioning of the Hypothalamic Pituitary Adrenal (HPA) axis (Mormède et al., 2007), although the concept of allostasis is being increasingly used in the context of animal welfare (Korte et al., 2007). However, the way in which an animal perceives the nature and intensity of a situation also affects its behavioural and physiological responses (Jones, 1997; Paul et al., 2005). Cognition refers to the mental abilities of animals; in particular their perception, reasoning and development of expectations. Consequently, cognitive ability and processing are major determinants of animals' reactions to different situations, not least the extent to which they are capable of experiencing suffering. The emerging areas of cognition and animal emotions are therefore increasingly important in animal welfare science (Boissy et al., 2007; Forkman et al., 2007; Jones and Boissy, 2011).

Of course there are many other approaches where the animal and its responses are the focus of attention. The cross-disciplinary and multidisciplinary approaches that are being increasingly adopted to address these advanced biological questions and the enhancement of animal welfare merit special mention. Examples of such collaborative ventures include the combination of a number of scientific disciplines with societal, ethical, economic and industry perspectives in international projects such as Welfare Quality®, the European Animal Welfare Platform, DIALREL, EUWelNet, etc.

There is also a considerable body of work and a wealth of knowledge regarding animal housing and husbandry. Early studies concentrated on comparing the effects on welfare of keeping animals in different housing systems under controlled conditions. This work produced useful results but it did not (and probably could not) reflect the wide range of housing systems found in practice. Therefore, research in animal welfare where the effects of housing and husbandry conditions are addressed frequently adopts an epidemiological approach that involves examination of animal physiology, behaviour, health and production under commercial conditions (e.g. Gunnarsson et al., 1999; Moinard et al., 2003). The importance of good stockmanship and good management has again come to the fore. For example, in the case of broiler chickens, it has been shown that environmental conditions in the house (humidity and ambient temperature in particular) are decisive factors (at least under the densities studied: 30-46 kg/m^2) that govern the birds' health and mortality (Dawkins et al., 2004). Since these environmental factors are determined by management practices there is a rapidly growing trend in animal welfare science towards developing decision support systems for farmers that can be used to reduce risks to welfare and to identify corrective actions whenever a problem is detected[4] (e.g. Bracke *et al.*, 2004; Jensen and Sorensen, 1998).

There are clear risk factors for poor welfare and recent research on animal welfare assessment has been an essential component of efforts to identify and control these risks. Many features of the animal itself (e.g. breed, age and reproductive stage), its housing (e.g. type of system and equipment) and its management (e.g. feeding routines and diets, handling) affect the risk of compromised welfare. In this way, welfare assessment can be regarded as measuring the 'output', in other words it can determine whether or not specific risks were actually realised in a particular case. By identifying problem areas it can also guide decisions on the most appropriate remedial strategies.

Current research is increasingly directed towards determining farmers' motivation to adopt more welfare friendly practices and how their motivation can vary according to the economic and other (e.g. labour conditions, time budgets) implications of implementing these changes. For example, it is already known that poor welfare can seriously damage the animals' performance and the farmers' profits (Hemsworth and Coleman, 1998; Jones, 1997). There is also mounting evidence from animal science that various welfare improvements can help to reduce costs or provide other economic benefits to farmers. For instance, providing pigs with more space will improve their growth rate (e.g. Edwards *et al.*, 1988; Gonyou *et al.*, 2005), handling dairy cows gently rather than roughly during milking can increase milk production (Rushen *et al.*, 1999), giving chickens frequent positive (or even neutral) human contact improved their food conversion ratios and decreased their fear of people (Gross and Siegel, 1979; Hemsworth and Coleman, 1998; Jones, 1995), and placing simple string devices in laying hen cages reduced feather pecking and the resultant feather damage in both experimental and commercial situations (Jones *et al.*, 2004.; McAdie *et al.*, 2005).

Ultimately an increased understanding of which indicators reliably reflect an individual animal's welfare status and of factors that may represent a risk to good welfare has to be implemented in practice to determine problems and guide solutions. This means that there is considerable pressure on animal scientists to deliver ways of measuring welfare for such implementation that are not only valid, reliable and robust but also feasible. Thus, a major contribution that animal welfare science can make in the context of the food chain is to provide a sound basis for practical welfare assessment and to assist in the process whereby our knowledge of animal welfare science is combined with that of food science for meaningful inclusion in European Union policy. However it is crucial to underline that animal welfare is a

---

[4] Decision support systems are also relevant to other important aspects in the food chain such as food safety and quality.

multidimensional concept that encompasses many aspects of the animal's life and society's evolving attitudes to animal welfare (i.e. what was considered acceptable in terms of animal suffering due to farming practices fifty years ago is nowadays no longer acceptable). Moreover, there are different values and preferences (both within animal science and the general public) regarding positive aspects of animals' lives that might lead to different courses of action, as pointed out by Fraser (2009):

> We see various [animal welfare] concerns that can be grouped roughly under three broad headings: one centres on the affective states of animals, one on the ability of animals to lead reasonably natural lives, and one emphasises basic health and functioning. These are not, of course, completely separate or mutually exclusive. For example, the advocates of natural living clearly believe that allowing animals to live a more natural life would make them more happy and healthy, and advocates of health and functioning clearly believe that unhealthy animals would suffer. Nonetheless, the different areas of emphasis are sufficiently independent that they can lead to quite different actions. A veterinarian may want animals kept individually in sterile cages to isolate them from disease, while an animal behaviour scientist wants an enriched group enclosure so that animals can behave more naturally, and both claim to be promoting animal welfare. Or an organic farmer may insist that animals have better welfare in a free-range system because it is more natural, while a confinement farmer insists that animals are better off indoors where they are protected from storms and predators. These are not, of course, factual disagreements: the veterinarian and the animal behaviourist, the organic farmer and the confinement producer, all may agree on the factual issues such as the mortality rate, whether the animals can perform certain behaviour and whether a particular veterinary procedure causes pain. The disagreement stems from different values – from different views about what is most important for animals to have a good life.

Thus the question of what is a good life for farmed animals generates some controversies and it is not one that can be solely addressed by animal science nor just by the perceptions and demands of the general public or farmers. Integration of results obtained from studies in the social and natural sciences has to be pursued if societal values and fundamental ethical issues are to be considered. Thus, workable science-based approaches must also address the concerns of the public (using the term in its widest interpretation to include farmers, processors, retailers, consumers, stakeholders, NGOs and so forth). To effectively consider and probe the issue of animal welfare we require not only a range of measures to identify and evaluate the numerous factors that can influence animals' lives, their wellbeing and performance but also a readiness to take account of the ethical concerns and growing expectations of the public. In order to address this challenge the Welfare Quality® project utilised a large consortium of both social and animal scientists that worked together towards the construction of the Welfare Quality® assessment protocols while simultaneously establishing a fruitful dialogue between science and society. This dialogue involved numerous interactions between animal scientists, social scientists, stakeholders and members of the public. It took several forms including some conventional methods of communication and exchange like: meetings, conferences, workshops, interviews, websites, newsletters and fact sheets, as well as some more experimental methods or hybrid forums (Callon *et al.*, 2009) like focus groups discussions, integration meetings and citizen and farmer juries (Miele *et al.*, 2011).

This chapter will now discuss three main issues in greater detail. First, we provide an overview of the policy instruments that the European Union has adopted (the body of EU regulations) in order to improve the welfare of farm animals; in many ways these underpinned the Welfare Quality® mission to harmonise the methods for assessing and monitoring the welfare of farm animals. Secondly, we report on farmers' initiatives in Europe for improving animal welfare. Thirdly, we describe our investigation of citizens' engagement with animal welfare issues in Europe and we discuss the conditions required for effective actions of consumers on the market, e.g. we address what in the literature is the widely recognised '*behavioural gap*' of European consumers between a reported high concern for animal welfare and a low purchasing of animal friendly labelled products (Kjærnes, 2012)

2.4 EU policy and legislation on the protection of farm animal welfare

European countries have a long history of concern and of legislation designed to protect the welfare of farm animals. The Council of Europe (founded in 1949 and now grown to include 47 members including EU member states) has three main conventions for the protection of farm animals which are intended to safeguard animals undergoing international transport (1968), those kept for farming purposes (1976) and those undergoing slaughter (1979). The EU is a signatory to each of these conventions. In addition, in the 1997 Treaty of Amsterdam (European Union, 1997), the EU introduced a Protocol on Protection and Welfare of Animals that revised the Treaty of Rome (the original Treaty setting up the EC). This Protocol is an important addition to EU policy and is legally binding for member states. It states that 'The High Contracting Parties, desiring to ensure improved protection and respect for the welfare of animals as sentient beings, have agreed upon the following provision which shall be annexed to the Treaty establishing the European Community, in formulating and implementing the Community's agriculture, transport, internal market and research policies, the Community and the Member States shall pay full regard to the welfare requirements of animals, while respecting the legislative or administrative provisions and customs of the Member States relating in particular to religious rites, cultural traditions and regional heritage'.

There are two important aspects to this Protocol. First is the recognition in EU law of animals as sentient beings (i.e. capable of feeling pleasure and pain). Second is the EU's stated desire to protect and pay full regard to the welfare of animals, albeit subject to many considerations within member states be they legal, administrative, cultural or religious.

There is a host of different policy instruments that governments and others can use to protect and improve the welfare of farm animals. These are briefly outlined and critically appraised in Bennett and Thompson (2011). Legislation has been an important policy tool within the EU with several pieces of legislation relating to farm animal welfare (Bennett and Appleby, 2010). Such legislation includes minimum welfare standards for animals kept for farming purposes (Directive 98/58/EC) and more specific EU farm animal welfare legislation which includes rules for:

1. Laying hens (Directive 99/74/EC, which specifies for instance minimum space requirements per hen and prohibits non-enriched cage systems by 2012).
2. Calves (Directive 91/629/EEC, amended in 1997, and 2008/119/EC which prohibits use of confined individual pens and muzzling of calves and also specifies minimum space requirements, feeding, etc.).
3. Pigs (Directive 2008/120/EC that requires that sows and gilts are kept in groups and gives standards for flooring, living space, access to rooting materials, training of stockmen as well as scientific advice for light requirements, noise levels, accommodation, materials for rooting and playing, access to fresh water, restrictions to mutilations and minimum weaning age).
4. Broiler chickens (Directive 2007/43/CE, which specifies maximum stocking densities and lays down conditions for lighting, litter, feeding and ventilation requirements).
5. Animal transport (e.g. Regulation EC 1/2005 on the protection of animals during transport and related operations).
6. Slaughter of animals (e.g. Directives 74/577/EC on stunning animals before slaughter, 93/119/EEC and Regulation EC 1099/2009 on the protection of animals at the time of killing, covering a wide range of animals and circumstances).

The European Commission puts forward proposals for protection of animal welfare that take account of the public concerns within the EU but which are based on scientific evidence and advice.

There are fundamental reasons why there is (and will always be) a need for government and/or EU intervention to protect and improve animal welfare. Firstly animal welfare is largely perceived as a 'public good' by European citizens (see Miele and Evans, 2010; Miele *et al.*, 2011) and even though there are livestock products that have animal welfare attributes in the eyes of some consumers it is not a widely marketed commodity in many European countries[5]. Secondly, we must also bear in mind that the production and consumption of farm animal products can give rise to 'animal welfare externalities' – benefits or costs that are incurred by people who are not involved in either the production or consumption of the animal products involved. For example, some people who are concerned about animal welfare and believe that keeping laying hens in cages causes those hens to suffer may avoid eating eggs from caged hens themselves but still experience negative feelings (a welfare loss to them) because they know that other people consume eggs from caged hens and thereby help perpetuate poor welfare for the birds. Other than making vocal or written protest there is little that such persons can do to reduce their own welfare loss since they can only make consumption decisions for themselves and not others. In addition, everyone else that shares these beliefs may also suffer a welfare loss. However, in cases where animal welfare is thought to be improved by other people's decisions on consumption they could experience a welfare gain; it is for this reason that animal welfare is often described as a 'public good' (see Bennett, 1995). In these circumstances, it could be argued that markets for animal products fail to adequately supply the level of animal welfare that society wants and that there is therefore a need for national governments or the EU to intervene to ensure that this failure is addressed and that both animal and societal welfare are improved.

It is in recognition of the need for market actors, governments and the EU to intervene in this way that many animal welfare charities and NGOs lobby not only industry players but also legislators to introduce various market and statutory measures to protect animal welfare. Indeed, animal welfare interests are relatively well represented by a diversity of different organisations ranging from charities and lobby groups (e.g. the World Society for Protection of Animals, Royal Society for the Prevention of Cruelty to Animals, Compassion in World Farming, Eurogroup for Animals, just to mention a few) to independent and government advisory bodies (e.g. the UK's Farm Animal Welfare Committee and a consortium of similar groups in other European countries which meet together as Euro-FAWC). Thus, for the reasons outlined above, animal welfare has an important political dimension within Europe. Despite the continuing need for more stringent EU regulation argued above, animal welfare legislation can be a relatively slow, blunt and costly instrument to apply, and its enforcement can be difficult as indicated by the delays and bottlenecks encountered in the implementation of recent Directives (e.g. abolition of non-enriched battery cages for laying hens)[6]. Increasingly, policy makers are seeking additional policy instruments to improve

---

[5] Probably the only exception are the products labelled 'Freedom Food' in the UK (see more in Chapter 3).

[6] Directive 1999/74/EC which introduces a ban on the use of un-enriched cages for laying hens. Belgium, Greece, Spain, France, Italy, Cyprus, Hungary, the Netherlands, Poland and Portugal in August 2012 still allow the use of un-enriched cages for laying hens despite the ban which came into force in January 2012 for which they have had 12 years to prepare.

Improving farm animal welfare

animal welfare, particularly the use of market mechanisms. So, although commodity markets, on their own, cannot be relied upon to provide the level of animal welfare that people in society want (for the reasons outlined above), some things can be done to improve the functioning and efficiency of markets. For example, for markets to function efficiently both buyers and sellers require adequate information but there is evidence to suggest that accurate information is sorely lacking in the case of farm animal welfare. Consumers who wish to purchase food animal products where the animals have been produced to high standards of animal welfare face a very difficult selection task in food stores (see Miele and Evans, 2010 for an example). It is almost impossible for such consumers to obtain clear and unbiased information on the levels of welfare enjoyed by animals in producing the food products that the consumer is considering purchasing. In most cases, there is no information on the animal welfare status of the food product and in other cases there are only rather oblique and somewhat vague references to characteristics of the good that consumers might associate with welfare, such as 'free-range'. A survey of consumers as part of the Welfare Quality® project found that only an average of 32% of consumers in Great Britain, Italy and Sweden felt sufficiently informed about animal welfare and that an average of 89% of respondents thought the inclusion of an animal welfare assurance mark and grading system on food to be important (Mayfield *et al.*, 2007). It seems clear, therefore, that many consumers would like more and better information about the animal welfare status of their food-producing animals and there is some evidence to suggest that a number of the more committed/ reflexive consumers might then act on this information and buy higher welfare foods. For example, the introduction of EU legislation for the mandatory labelling of eggs produced from caged hens resulted in a substantial increase in the purchase of free-range and barn eggs[7]. On the other hand, many consumers fail to translate their desire for higher levels of farm animal welfare into purchase choices and would prefer to be reassured (not only by legislators but also by supermarket chains) that all animal products available in ordinary shops are obtained at an acceptable level of animal welfare (Evans and Miele, 2012; Miele and Evans, 2010). Of course, cost is also an issue that cannot be ignored, particularly in times of global economic difficulty.

The development of a welfare labelling system for livestock products has been identified for some time as a desirable approach by the European Commission and

---

[7] Defra announced that in year 2011, 51% of the nine billion eggs laid in the UK came from barn, free-range or organic hens – marking the tipping point in a quiet revolution in shopping habits since 1995, when 86% of eggs came from battery cages. Many retailers such as the Co-op and Marks & Spencers have stopped selling eggs from caged birds altogether. This buying trend is predicted to continue, with 52% of eggs from barn, free-range or organic sources set to be consumed in 2012. Experts believe the shift in buying habits was triggered by the introduction of compulsory labelling eight years ago, which forced producers to state the method of production. Available at: http://www.scotsman.com/the-scotsman/personal-finance/uk_sales_of_free_range_eggs_outsell_battery_production_1_2123975.

this aim influenced the mission of the Welfare Quality® project. Clear information for consumers on the animal welfare status of food-producing animals backed by scientifically-based, certified welfare assessments will allow consumers to make informed and better decisions regarding their food purchases and thereby satisfy their preferences for animal welfare. In principle, this should mean that both people and the animals are better off as a result.

But it is not only consumers that would benefit from better information about animal welfare. Livestock producers are themselves often unsure about the levels of welfare experienced by the animals in their care for a number of reasons; these range from insufficient record keeping, lack of awareness of the welfare implications of certain conditions or methods of keeping animals, conditions which have negative influences on welfare are often unrecognised or undiagnosed, and there may be no or little formal welfare assessment at any time during the lifetime of the animal. The Welfare Quality® protocols for the assessment of farm animal welfare address this issue and provide farmers with very useful information they can use to not only improve the welfare of the animals in their care but also to demonstrate the resultant 'better welfare'/ improved quality of life to other producers. As indicated above, improved welfare status can also provide economic benefits for their businesses in terms of increased productivity, product quality and profitability (Hemsworth, 2003; Jones and Boissy, 2011; Stott *et al.*, 2005; Waiblinger *et al.*, 2006).

The Commission also supports and initiates international initiatives with third countries to create greater consensus and raise awareness on animal welfare (European Commission, 2007, 2012). This clearly responds to common trends that animal welfare issues are globalising. Animal welfare is, for example, increasingly a topic of discussion in emerging economies in Asia (including China) and Latin America. International institutions such as OIE (World Organization for Animal Health), FAO (Food and Agriculture Organization) and the IFC (International Finance Corporation, which is a member of the World Bank Group) are increasingly active in developing animal welfare policies (Ingenbleek and Immink, 2011).

In this context, the European Commission also works with international organisations such as the OIE towards the promotion, development and implementation of internationally recognised animal welfare standards.

2.5 Changes in farming and in stakeholders' concerns for animal welfare

Animal farmers in Europe find themselves right in the middle of a heated public debate about animal welfare (Bock, 2009). Farming has often been represented by certain actors as exploitation because it simply involves producing animals for slaughter (Lund and Olsson, 2006). However, numerous studies point to the fact that farmers do care about their animals and feel empathy or even affection for them. Indeed, many of them believe that caring about the animals is an important element of farmers' identity and culture (Porcher, 2006; Tovey, 2003) and a crucial characteristic of being a 'good farmer' (Bock and Van Huik, 2006, 2007; Bock et al., 2007; Borgen and Skarstad, 2007; Bruckmeier and Prutzer, 2007; De Greef et al., 2006; Dockes and Kling Eveillard, 2006; Lassen et al., 2006). They also know that society expects them to look after their animals with diligence and care and that the use and killing of animals becomes acceptable only when farmers demonstrate attachment to their animal and when animals have a 'good life' before their death. Previous research into the relationship between farmers and their livestock revealed how human-animal interaction affects the welfare and productivity of animals (Hemsworth, 2003; Hemsworth and Coleman, 1998; Hemsworth et al., 2000; Jones, 1997; Lensink et al., 2001). These studies demonstrated how good care improved the wellbeing of the animals in terms of health and productivity. But it also works the other way round: having good relations with animals lightens the burden of farm work and renders some parts of the work less dangerous and less stressful for both the animals and the farmers. One might argue, then, that a 'good' and successful farmer really needs to care about his or her animals. Yet questions remain about whether all farmers care the same for their animals and about the extent to which different sectors and husbandry systems influence this. In the following section we provide an overview of the commonalities and differences in the attitudes of animal farmers, based on recent research carried out as part of the Welfare Quality® project. The research focused on farmers working in the pig, cattle and poultry sectors in seven European countries: Italy, France, United Kingdom, Norway, Sweden, the Netherlands, and Hungary (Bock and Van Huik, 2007, 2008, 2009; Bock et al., 2007, 2010a,b; Roe et al., 2009).

It became clear that most farmers regard looking after their animals and ensuring their wellbeing as an essential aspect of their job. In interviews carried out by Welfare Quality® researchers many farmers emphasised the need to satisfy the physiological needs of animals, including their health. They are also concerned with psychological aspects of well-being such as the absence of stress and the importance of good relationships with the humans who take care of them. They keep a close eye on the wellbeing of the animals by checking their external appearance (e.g. skin, eyes, posture), observing their behaviour (e.g. playfulness, social interaction, mobility) and checking their performance (Bock et al., 2007; Dockès and Kling-Eveillaard, 2006; Hubbard and Scott, 2011). Taking good care of the animals constitutes an important component of the farmers' professional pride and ethics, and they would very much like to see more recognition of their professional knowledge of and engagement with animal welfare. Most farmers realise that many members of the public have concerns about farm animal welfare and that these concerns need to be addressed in order for them to keep their licence to produce (Kjærnes *et al.*, 2009). They worry, however, about the public's lack of knowledge about animal farming and biology as well as the appropriateness of the definition(s) of animal welfare used in the public debate, in which they often feel unjustly stigmatised (Bock, 2009).

Overall, our research shows that farmers demonstrate considerable readiness to accept more stringent animal welfare regulations as long as these allow them to remain in business. They worry, however, that consumers and retailers may not be ready to pay the higher price that would be required to cover the extra production costs that the farmers expect to result from additional and/or more stringent welfare requirements. Farmers also fear that they could be side-lined in the global market where producers from outside the European Union may not be obliged to comply with the same regulations (Kjærnes *et al.*, 2009).

Notwithstanding these commonalities, the research also demonstrated systematic variations in farmers' views and concerns related to differences between husbandry sectors and production systems, assurance schemes, and countries.

2.5.1 Differences between sector: management and regulation

Our research was carried out in the following sectors: pigs (breeding and fattening), cows (dairy, beef and veal) and chickens (eggs and meat), which differ considerably in how animals are kept and treated. They are also quite differently regulated by national and European laws as well as private certification schemes. Generally speaking, the more intensive production sectors are the most regulated by legislation and private standards (Bock and van Leeuwen, 2005).

Farmers engaged in intensive production sectors, such as for chickens, pigs and calves kept for veal production, are generally more aware of the need to address animal welfare concerns at farm level. Compared to farmers keeping dairy and suckling cows, they have more elaborate ideas about what animal welfare is, are more aware of regulations and the existence of assurance schemes and, on average, are more interested in participating in assurance schemes (Kjærnes *et al.*, 2009). In part this may be explained by the fact that farmers in these intensive (sub)sectors have already experienced tightening regulations and increasing public debate. It is also in these sectors that farmers have little choice but to enter assurance schemes in order to achieve market access (Bock and Van Huik, 2009; Bock *et al.*, 2007).

The differences between the above mentioned more or less intensive (sub)sectors are not only related to differences in regulation. They have also to do with management structures, influencing the ways in which animals are kept on farms and the levels of contact between animal and farmer (Bock *et al.*, 2007; Miele and Bock, 2007). To give an example: dairy farmers encounter animals closely during milking every day, while bull-fattening farmers keep at a distance for their own safety. The type of production and housing system also determines how many animals a farmer cares for and for how long an animal is actually present at the farm. Dairy cows, for instance, stay in the same farm for several years whereas a veal calf is sold within months. Generally speaking, farmers are more concerned about the welfare of those animals that are kept at their farm for longer periods and when caring for them involves regular close contact. At the same time, farmers tend to defend their own husbandry system, underlining its advantages in terms of animal welfare. For example, farmers who keep their hens in cages often point out that the birds are better protected from disease, predation, cannibalism, etc. than those which are kept outdoors.

2.5.2 Basic and high standard quality assurance schemes versus animal welfare and organic schemes

Differences of opinion are also found among farmers about the various types of Quality Assurance Schemes (QAS). In our studies we distinguish between basic and top farm assurance schemes and specific animal welfare and organic schemes (Bock and Van Leeuwen, 2005). Most notable is the difference between farmers engaged in specific animal welfare and organic schemes on the one hand and farmers in no, basic and top quality schemes on the other. The two groups differ with regard to their definition of animal welfare, their readiness to accept more stringent animal welfare regulations, and their belief in the possibility to market animal welfare by way of labelling 'animal friendly' products (Kjærnes *et al.*, 2008, 2009). Most farmers engaged in no, basic or top quality assessment schemes define animal welfare primarily in terms of the animal's health and zoo-technical performance. Animals fare well in their view when their biological needs are sufficiently met and sickness, pain and injuries, as well as stress, are prevented or minimised. When animal welfare is assured in these terms, animals are healthy and they grow satisfactorily from the perspective of the farmer. Taking good care of the animals' welfare therefore makes sense economically and is part and parcel of being a good and successful farmer. On the other hand farmers engaged in specific animal welfare or organic schemes consider the animal's opportunity to express natural behaviours as crucial for its welfare, in addition to ensuring physical health and the fulfilment of basic biological needs. Taking care of the animals' welfare is important for them as part of their personal ethics and professional philosophy, but also for economic reasons.

Underlying the different attitudes of the two groups of farmers towards animal welfare issues is a more fundamental difference in farming style or production logic (Kjærnes *et al.*, 2009). The majority of the first group of farmers produces for the conventional market where the price is low and profit depends on selling large quantities of meat and on cost reduction. In this context a good farmer is an efficient farmer and a definition of animal welfare in terms of animal health and productivity makes economic sense. When the objective of agriculture is more broadly defined and includes issues like care for nature and the environment, such as in organic schemes, the definition of good farming and good animal welfare will concomitantly change as well. In this context 'naturalness' and natural behaviour are considered more valuable by the market-parties (consumers and retailers) as well as the farmer (Miele and Kjærnes, 2009). Specific animal welfare schemes and organic farming provide a context where such farming practice is stimulated and rewarded by a premium price that compensates for the higher costs involved (Bock and Van Leeuwen, 2005; Hubbard *et al.*, 2007). Farmers in no, basic and top QAS's are on the contrary obliged to increase production in order to make up for the costs resulting from more stringent animal welfare regulations. They often have to make investments that in their view add nothing to animal welfare and do not increase the economic value of their products. The markets where their products are sold are generally not ready to reward their engagement with premium prices (Bock *et al.*, 2007; Bock and Van Huik, 2007).

2.5.3 Differences between countries: global competition and regulation via market or state

There are major national differences in the size and organisation of animal farming, and its position on the world market. The opportunity to sell products and to cope with foreign competition depends, among other factors, on the openness of the national market and the level of competition between domestic and imported products. All farmers considered it unfair if they had to follow stringent regulations but needed to compete with others who could produce more cheaply under lighter regulations. Therefore farmers across all countries stressed the need for equal regulation. These issues of a level playing field were much more of a worry for pig and poultry farmers than for cattle farmers. Especially in countries that import a lot of meat from abroad, farmers consider the resulting competition as unfair (Hubbard *et al.*, 2007; Huik and Bock, 2007).

Political differences prove to be important as well. The different organisation and regulation of farmed animal welfare, either by way of the market or of the state, influences the frame of reference of farmers, their attitudes and perceptions and ultimately their perceived opportunities and choices (Kjærnes *et al.*, 2009). The latter aspect is also influenced by the relative heat of the public debate. Especially in the Netherlands and the United Kingdom animal farmers feel under pressure because of NGO's persistence and outspokenness in promoting the welfare of the animals.

Generally speaking, Norwegian and Swedish farmers are opposed to the idea of improving animal welfare by way of quality assurance schemes or labelling because they do not agree with the idea of market differentiation (Skarstad *et al.*, 2008; Terragni and Torjusen, 2007). Hitherto, there are few private assurance schemes in Norway and Sweden, perhaps because they have a tradition of institutionalising animal welfare by law. On the contrary, the Netherlands and the United Kingdom are increasingly regulating farm animal welfare by way of the market, which does offer some farmers the opportunity for commercial distinction. Indeed, for Dutch and UK farmers this 'high welfare' approach was perceived to be the most 'logical' way to move forward, although they were worried about the profitability of such a niche market. As mentioned above, they feared that the tightening of legislation would render even this market more difficult to compete in against increasing non-European production (Bock *et al.*, 2010b; Kjærnes *et al.*, 2009). In France and Italy there is so far little market demand for animal friendly meat but both countries have several private schemes oriented towards origin-based high quality production. As a result interest among French and Italian farmers in 'animal-friendly' production is increasing when linked to the notion and marketing of high quality products (Souquet *et al.*, 2006; Menghi, 2007; Kling-Eveillard *et al.*, 2009).

2.6 Public perception of animal welfare in Europe

Some commentators have recently pointed out that in the last decades a 'crisis of consumer confidence' in the European animal farming industry has emerged (Aslet, 2007) and many studies have shown that European citizens are concerned about farm animal welfare (Ingenbleek and Immink, 2011; Kjærnes, 2012). The extent of the public concern is evident in two recent Eurobarometer surveys of public attitudes towards farm animal welfare (carried out in 2005 and repeated in 2006)[8]. The vast majority of EU citizens interviewed in these large public consultation exercises declared that they were concerned about farm animal welfare. More specifically concerns were expressed about the living conditions of laying hens in battery cages, about problems associated with rapid growth of broiler chickens and about the welfare of pigs. Surprisingly, on the other hand, the welfare of dairy cows was not considered to be a problem. A significant minority of people, predominantly those living in old rather than new member states, declared that they thought about animal welfare while shopping for meat and other animal products and a high percentage of these people declared that they were willing to pay a (small) premium for animal products obtained in a welfare-friendly way. A high proportion of those interviewed (62%) declared that they would be willing to change where they shopped in order to find animal friendly products. They also mentioned the lack of availability of animal friendly products in ordinary shops and many lamented the lack of information provided on current food labels. However, most respondents also believed that the welfare of farm animals was better

---

[8] Eurobarometer, 2005 and Eurobarometer, 2007.

in the EU than elsewhere. Although some of these statistics seem contradictory, they might indicate awareness not only of the crises that have occurred in the animal farming population (swine fever, BSE, foot-and-mouth disease and avian influenza just to mention the most important ones in the last thirty years) but also of the recent initiatives taken by the EU, and by several national governments in Europe, to address these problems. For example, relatively new institutions, such as the Food Standards Agency (FSA) in the UK and the European Food Safety Authority (EFSA)[9] have been created to assess the safety of food and the risks associated with technological innovations in the food sector, to provide a forum for the expression of consumers' interests, and to promote research for evaluating the benefits that consumers could gain from food innovations (all of which may help to prevent future crises in consumer confidence)[10].

Within the Welfare Quality® project was carried out the largest in depth study of public perception of animal welfare in Europe to date (see Evans and Miele, 2007a,b; Kjærnes and Lavik, 2008; Kjærnes *et al.*, 2007). The investigation of the public's concerns about animal welfare included a series of focus group discussions with 'consumers' of animal products in seven European countries (Sweden, Norway, Italy, France, the Netherlands, the United Kingdom and Hungary), a large telephone survey of a sample of citizens in the same countries, and, in conclusion, a series of 'Citizen Juries' in three of the seven European study countries (Miele *et al.*, 2009, 2011).

Findings from the focus group research support the results of previous analyses in that knowledge or perceptions (Verbeke, 2009) about farm animal welfare tend to be shaped by a bipolar understanding of farming systems. 'Industrial' systems were perceived to provide low animal welfare whereas alternatives, such as 'organic' systems, were perceived to provide good welfare. Furthermore, participants tended to know more about those welfare issues that were perceived to be connected to food quality and safety (such as the use of antibiotics, animals' feed and levels of stress) than those which were perceived as having little influence on the final food product. Participants also lacked detailed technical knowledge about issues such as the nature of modern farming systems, the types of animal breeds used in modern intensive systems and issues surrounding farmed animal biology and physiology. This lack of understanding

---

[9] In the UK the Food Standards Agency is an independent Government department set up by an Act of Parliament in 2000 to protect the public's health and consumer interests in relation to food. See http://www.food.gov.uk/aboutus/how_we_work/originfsa for a mission statement and details on this organisation. The European Food Safety Authority (EFSA) was set up in January 2002, following a series of food crises in the late 1990s, as an independent source of scientific advice and communication on risks associated with the food chain. EFSA was created as part of a comprehensive programme to improve EU food safety, ensure a high level of consumer protection and restore and maintain confidence in the EU food supply (http://www.efsa.europa.eu/EFSA/efsa_locale-1178620753812_AboutEfsa.htm).

[10] For a discussion on issues of trust in the food industry in Europe see Kjarnes *et al.*, 2007.

Improving farm animal welfare

or *illiteracy* about modern farming systems does not prevent the majority of citizens from being concerned about the life of farm animals, on the contrary it is generating widely spread suspicion that modern, intensive animal farming systems are inherently un-friendly if not explicitly cruel to animals and that extensive, small scale, traditional, free range, and organic systems will automatically deliver better welfare.

Responses to survey questions as well as our focus group discussions indicate that the overwhelming majority of people is interested in animal welfare issues and that the public would like to be better informed about the welfare status of the animals used for producing the food they buy. However the majority of people who participated in the focus group discussion and the survey did not think that their own choices on the market (either boycotting products perceived as non-animal friendly or actively choosing animal friendly products) would be required (because they thought that the EU and the national governments were responsible for ensuring that a good level of animal welfare is achieved in Europe) or they thought that their own food choices would not make a significant difference, given the limited purchasing power of individual households. With the exception of Hungary, the majority tended to believe that farm animal welfare conditions in their own country had improved in the last few years and there was a widely held belief that European regulation was sufficiently stringent to guarantee a decent life for farmed animals. Most participants were interested in learning more about how food is produced and about the living conditions experienced by farmed animals. However, they did not necessarily want to be presented with detailed information about different production systems whilst shopping for food. Group discussions reflect that many considered farm animal welfare to be more of a 'public good' matter rather than an issue that should be addressed by the market. Analyses of the survey data indicate that the widespread interest in having more information on farmed animal welfare is not associated with any clear intention to act by changing shopping practices. The general assumption seems to be that the legal system and existing regulation within the food chain (both in the form of direct State regulation and supply chain actor governance through food standards and assurance schemes) does/will ensure that what is available on the market is produced to an acceptable ethical standard. Even though there is widespread scepticism towards many innovations in livestock production (e.g. use of antibiotics and growth promoters in animal feed, breed selection, cloning and so forth) and about some animal housing systems (e.g. all year indoor, laying hen cages), trust in the national and supranational governments to promote more stringent rules and, especially, in the capacity of the animal welfare NGOs to keep animal welfare high on the political agenda seems to be relatively high in a number of countries (e.g. Sweden, Norway, the United Kingdom).

In relation to their own shopping practices, only a minority of participants indicated a willingness to actively search for 'animal-friendly' products. Taste, quality, price, safety, convenience, and freshness were the most common factors taken into account when shopping for food. Given that the most widely voiced ethical concern was animal suffering or cruelty towards animals, the most common expectation was that any product that arrives on the market is obtained through methods of production that avoid animal suffering.

Many participants seemed to have very broad and general ideas about what a 'welfare-friendly purchase' is. The most generalised attitude is an overall approval of production systems and/or regulations in their own country. A significant proportion of participants associated organic, high-quality, and even simply locally-produced products with higher levels of animal welfare. The focus on production systems is also reflected in the popularity of free-range systems for laying hens' eggs and for chicken meat. There is a clear differentiation in welfare concerns between species; chicken farming is perceived as the most industrialised and intensive, resulting in poor welfare, followed next by pig farming, and then farming of cows for their milk. Higher welfare is associated with extensive production systems, especially those for niches of typical breeds renowned in a specific country, for example, in Norway national sheep production, in Hungary Mangalica pigs and grey cows, in Italy Cinta Senese pigs and so forth. While the public perception of animal welfare risks associated with many systems of production coincided with the 'experts' assessment, it seems that the welfare risks that extensively managed ruminants face (in terms of disease risk, climatic extremes, variable food supply, predation, etc. (Turner and Dwyer, 2007) are most often overlooked by the public.

There is, nevertheless, a gradual shift in societal attitudes and a growing awareness of a range of welfare issues. This is partly prompted by the activities of animal welfare organisations and by recent food scares, but also by the dramatic growth of quality labelling, notably organic. For some people, especially in Northern Europe, their search for 'animal friendly' food is associated with a more critical and active consumer role; indeed we found a minority of 'critical consumers' who actively search for 'animal-friendly' products. This group has grown in the last 10-15 years, especially in the United Kingdom and the Netherlands. Critical consumers seem to be willing to pay a higher price for 'welfare-friendly' products. By higher farm animal welfare they understand more natural systems of production (e.g. free range, organic and 'traditional' farms) and production systems that enable farmed animals to express more natural behaviours (e.g. dust bathing and nest building for hens, rooting for pigs). In contrast, they believe that issues such as the absence of pain or other negative emotions (such as fear and stress) are not sufficient to merit rewarding with higher purchasing prices, but rather represent the bare minimum quality of life that should be guaranteed to all animals reared as food or food sources. Despite the above findings, specific 'animal-friendly' products still represent only a small segment of the food market. It has become increasingly clear that public concern for animal welfare is not matched by consumers' willingness to pay for it (Evans and Miele, 2007b; IGD, 2007). Another finding is the above mentioned preference of consumers to consider farm animal welfare a public good that is (and it should be) regulated by the public authorities and taken care of by the producers and retailers rather than facing choices that they may find upsetting (SEERAD, 2003).

Moreover there is a relatively common association of higher farm animal welfare with product quality. Participants in focus group discussions had high expectations about the 'animal friendliness' of a range of 'quality' products that were currently available. This was also reflected in the telephone survey results. This association of 'welfare friendliness' with food quality is especially prominent in France, but is also visible in many other countries. Quality products (for the most part those that are thought of as having a superior taste) are widely assumed to result from production processes that exhibit higher levels of welfare. Often a link is also made between healthy products and increased levels of welfare. The idea of 'good for animals, good for humans' is widespread everywhere, but seems to be particularly emphasised in Italy, Hungary, and France. Of course, there is also scientific evidence of a positive relationship between welfare status and product quality (Hemsworth, 2003; Jones and Boissy, 2011).

Our research findings point to the need to address public concern about farm animal welfare through a range of different mechanisms. A key issue is the trustworthiness of the information that is provided regarding the welfare status of farmed animals. Undoubtedly, improving the *transparency* of the market through regulated labelling and better communication to EU citizens about farm animal welfare issues would help to create a suitable climate for more consumers to become actively engaged and to be able to translate this engagement into informed choices[11]. However, for the market mechanism to work, more transparency about the welfare status of products should be coupled with more *accountability* of the animal supply chains when welfare claims are not explicit on the products and more *commensurability* of the existing claims on animal products. This points to the importance of efficient monitoring and sanction systems, performed by independent parties.

However the majority of EU citizens also expect public intervention, such as stringent regulation on key aspects of animals' lives like transport and slaughter, as well as

---

[11] Even though there are clear indications that a large proportion of the European public is not inclined to accept the 'marketization of animal welfare (or equipped for dealing with it), for a discussion on this point see Miele and Evans (2010).

the provision of financial and educational support for farmers, raising minimum standards by legal means, and banning the most problematic systems of production. All of these interventions are perceived by the European public as a priority to improve the quality of life for the majority of farmed animals in Europe.

2.7 Conclusion

This chapter has introduced the starting points, ethos and key elements of the Welfare Quality® project (the project's vision is described in Chapter 4). At a time in which many member states of the European Union are facing increasing difficulties in implementing the current EU regulations on animal welfare, the EU Commission has begun a process, with the Animal Welfare action plan 2006-2010 and the EU Animal Welfare Strategy 2012-2015, for identifying new instruments, including the harmonisation of animal welfare claims on animal products in the market for increasing market transparency and facilitating informed decision making and actions by consumers. This market mechanism, if coupled with existing and more stringent future regulation on animal farming, could create better opportunities for the emergence of innovations for animal friendly farming systems. By developing an assessment and information tool that would allow the harmonisation of welfare claims on animal products[12] the Welfare Quality® project directly addressed this requirement. This tool is soundly based on the developments in animal welfare science as well as insights into initiatives taken by key stakeholders (farmers, retailers, NGOs, etc.) and the public's concerns towards animal welfare.

It is recognised that the farming of animals is no longer viewed by European citizens simply as a means of food production. Instead, it is seen as fundamental to other key social goals such as food safety and quality, environmental protection, sustainability, and enhancing the quality of life in rural areas (Gavinelli *et al.*, 2007). Therefore it seems highly likely that animal welfare within Europe will continue to increase in importance with an increasing number of policy initiatives designed to protect and improve the welfare of animals that involve a mix of policy instruments, legislation and market-oriented mechanisms. It is hoped that the outputs from Welfare Quality® and their further development, with the activities of the European Animal Welfare Platform and the Welfare Quality Network as outlined in later chapters, will greatly help policy makers, the industry and the others constituencies in society that perceive themselves as involved in this regard.

---

[12] At least in the three species (cattle, poultry and pigs) and 7 production types studied.

References

Aslet, C. (2007). Bird flu is the price of your £5 roast, The Observer, 4 February 2007, 27.

Bennett, R. and Appleby, M. (2010). Animal welfare in the European Union. In: Oskam, A., Meester, G. and Silvis, H. (eds.) EU policy for agriculture, food and rural areas, Wageningen Academic Publishers, Wageningen, the Netherlands, pp. 243-251.

Bennett, R. and Thompson, P. (2011). Economics. In: Appleby, M.C, Mench, J.A., Olsson, I.A.S. and Hughes, B.O. (eds.) Animal welfare. 2nd ed. CAB International, Wallingford, UK, pp. 279-290.

Bennett, R.M. (1995). The value of farm animal welfare. Journal of Agricultural Economics, 46(1), 46-50.

Blokhuis, H.J. (1999). Integration of animal welfare in intensive animal production. In: Wensing, Th. (ed.) Production diseases in farm animals, Wageningen Pers, Wageningen, the Netherlands, pp. 222-229.

Blokhuis, H.J., Hopster, H. Geverink, N.A. Korte, S.M. and Van Reenen, C.G. (1998). Studies of stress in farm animals. Comparative Haematology International, 8, 94-101.

Bock, B.B. (2009). Farmers' perspectives. In: Butterworth, A., Blokhuis, H.J., Jones, B. and Veissier, I. (eds.) Proceedings conference, delivering animal welfare and quality: transparency in the food production chain including the final results of the Welfare Quality® project, 8-9 October 2009, Uppsala, Sweden, pp. 73-75.

Bock, B.B. and Van Huik, M.M. (2007). Animal welfare, attitudes and behaviour of pig farmers across Europe, British Food Journal, 109(11), 931-944.

Bock, B.B. and Van Huik, M.M. (2008). Attitudes, beliefs and behaviour of cattle farmers. In: Kjærnes, U., Bock, B.B., Higgin, M. and Roex, J. (eds.) Farm animal welfare concerns; consumers, retailers and producers, Welfare Quality® Reports no. 7, Cardiff University, Cardiff, UK, pp. 257-319.

Bock, B.B. and Van Huik, M.M. (2009). Attitudes, beliefs and behaviour of poultry farmers across Europe. In: Kjærnes, U., Bock, B.B., Higgin, M. and Roex, J. (eds.) Farm animal welfare within the supply chain: regulation, agriculture, and geography, Welfare Quality® Report no. 8, Cardiff University, Cardiff, UK, pp. 77-144.

Bock, B.B. and Van Leeuwen, F. (2005). Socio-political and market developments of animal welfare schemes. In: Roex, J. and Miele, M. (eds.) Farm animal welfare concerns: consumers, retailers and producers. Welfare Quality® Reports no 1. Cardiff University, Cardiff, UK, pp. 115-167.

Bock, B.B., Swagemakers, P., Jacobsen, E. and Ferrari, P. (2010b). Part IV Farmers' juries: a synthesis. In: Bock, B.B., Swagemakers, P., Jacobsen, E. and Ferrari, P. (eds.) Dialogue between farmers and experts regarding farm animal welfare; farmers' juries in Norway, the Netherlands and Italy, Welfare Quality® Reports no. 17, Cardiff University, Cardiff, UK, pp. 97-131.

Bock, B.B., Swagemakers, P., Lever, J., Montanari, C. and Ferrari, P. (2010a). Farmers' experiences of the farm assessment: interviews with farmers. In: Bock, B.B. and De Jong, I. (eds.) The assessment of animal welfare on broiler farms, Welfare Quality® Reports no. 18, Cardiff University, Cardiff, UK, pp 29-95.

Bock, B.B., Van Huik, M.M., Prutzer, M., Kling Eveillard, F. and Dockes, A. (2007). Farmers' relationship with different animals: the importance of getting close to the animals. Case studies of French, Swedish and Dutch cattle, pig and poultry farmers. International Journal of Sociology of Agriculture and Food, 15(2), 108-125.

Boissy, A., Manteuffel, G., Jensen, M.B., Moe, R.O., Spruijt, B., Keeling, L.J., Winckler, C., Forkman, B., Dimitrov, I., Langbein, J., Veissier, I. and Aubert, A. (2007). Assessment of positive emotions in animals to improve their welfare. Physiology and Behaviour, 92, 375-397.

Borgen, S.O. and Skarstad, G.A. (2007). Norwegian pig farmers' motivations for improving animal welfare. British Food Journal, 109(11), 891-905.

Bracke, M.B.M., Hulsegge, B., Keeling, L. and Blokhuis, H.J. (2004). Decision support system with semantic model to assess the risk of tail biting in pigs: 1. Modelling. Applied Animal Behaviour Science, 87, 31-44.

Brambell Committee (1965). Report of the technical committee to enquire into the welfare of animals kept under intensive livestock husbandry systems. Command Report 2836. London: Her Majesty's Stationary Office.

Broom, D.M. (1996). Animal welfare defined in terms of attempt to cope with the environment. Acta Agriculturae Scandinavica (Section A – Animal Science), 27, 22-29.

Bruckmeier, K. and Prutzer, M. (2007). Swedish pig producers and their perspectives on animal welfare – a case study. British Food Journal, 109(11), 906-918.

Bruinsma, J. (2003) World agriculture: towards 2015/2030, an FAO perspective. Earthscan Publications, London, UK.

Buller, H. and Cesar, C. (2007). Eating well, eating fare: farm animal welfare in France. International Journal of Sociology of Food and Agriculture, 15 (3), 45-58.

CBS, 1998. Statistisch jaarboek 1998. Centraal Bureau voor de Statistiek, Voorburg/Heerlen, the Netherlands.

Dawkins, M.S., Donnelly, C.A. and Jones, T.A. (2004). Chicken welfare is influenced more by housing conditions than by stocking density. Nature, 427, 342-344.

De Greef, K., Stafleu, F. and De Lauwere, C. (2006). A simple value distinction approach aids transparency in farm animal welfare debate. Journal of Agricultural and Environmental Ethics, 19, 57-66.

Dockès, A.C. and Kling Eveillard, F. (2006). Farmers' and advisers' representations of animal and animal welfare. Livestock Science, 103, 243-249.

Duncan, I.J.H. (1993). Welfare is to do with what animal's feel. Journal of Agricultural Environment and Ethics, 6, 8-14, [Supplement].

Edwards, S.A., Armsby, A.W. and Spechter, H.H. (1988). Effects of floor area allowance on performance of growing pigs kept on fully slatted floors. Animal Production, 46, 453-459.

European Commission (2007). Attitudes of EU citizen towards animal welfare. Special Eurobarometer 270, Wave tt.1, TNS Opinion & Social. European Commission, Brussels, Belgium.

European Commission (2012), Communication from the commission to the European Parliament, the Council and the European Economic and Social Committee on the European Union Strategy for the Protection and Welfare of Animals 2012-2015 (Text with EEA relevance), {SEC(2012) 55 final}, {SEC(2012) 56 final}, 15.2.2012 COM(2012) 6 final/2, European Commission, Brussels, Belgium.

European Union (1997). Treaty of Amsterdam. Office for Official Publications of the European Communities, Luxembourg, Luxembourg.

Evans, A. and Miele, M (2007a). Consumers' views about farm animal welfare, Part 2: Comparative report based on Focus Group Research. Welfare Quality® Reports no. 5, Cardiff University, Cardiff, UK.

Evans, A. and Miele, M. (2012). Between food and flesh: how animals are made to matter (and not to matter) within food consumption practices, Environment and Planning D – Society and Space, 30(2), 298-314.

Evans, A. and Miele, M. (eds) (2007b). Consumers' views about farm animal welfare, part I: national reports based on focus group research, Welfare Quality® Reports no. 4, Cardiff University, Cardiff, UK.

FAO (2004) FAOSTAT on-line statistical service. FAO, Rome, Italy. Available at: http://apps.fao.org.

Forkman, B., Boissy, A., Meunier-Salaün, M.-C., Canali, E. and Jones, R.B. (2007). A critical review of fear tests used on cattle, pigs, sheep, poultry and horses. Physiology and Behavior, 92, 340-374.

Fraser, D. (2005). Animal welfare and the intensification of animal production. An alternative interpretation. Food and Agriculture Organization of the United Nations, Rome, Italy, 28 pp.

Fraser, D. (2008). Understanding animal welfare: the science in its cultural context. Wiley-Blackwell, London, UK.

Fraser, D. (2009). Assessing animal welfare: different philosophies, different scientific approaches. Zoo Biology 28, 507-518.

Gavinelli, A., Rhein, C. and Ferrara, M. (2007). European policies on animal welfare and their effects on global trade. Farm Policy Journal, 4(4), 11-21.

Gonyou, H.W., Deen, J., McGlone, J.J., Sundberg, P.L., Brumm, M.C., Spoolder, H., Kliebenstein, J., Buhr, B. and Johnson, A.K. (2005). Developing a model to determine floor space requirements for pigs. Journal of Animal Science, 82(Suppl. 2), 34.

Grant, W. (1997). The common agricultural policy. Macmillan, London, UK.

Gregory, N. (1998). Physiology of stress, distress, stunning and slaughter. In: Gregory, N. (ed.) Animal welfare and meat science. CABI, Wallingford, UK, pp. 64-92.

Gross, W.B. and Siegel, P.B. (1979). Adaptations of chickens to their handler and experimental results. Avian Diseases, 23, 708-714.

Gunnarsson, S., Keeling, L.J. and Svedberg, J. (1999). Effect of rearing factors on the prevalence of mislaid eggs, cloacal cannibalism and feather pecking in commercial flocks of loose housed laying hens. British Poultry Science, 40, 12-18.

Harrison, R. (1964). Animal machines. Vincent Stuart Ltd., London, UK, pp. 186.

Hemsworth, P.H. (2003). Human-animal interactions in livestock production. Applied Animal Behaviour Science, 81, 185-198.

Hemsworth, P.H. and Coleman, G.J. (1998). Human-livestock interactions: the stockperson and the productivity and welfare of intensely farmed animals. CAB International, Wallingford, UK.

Hemsworth, P.H., Coleman, G.J. Barnett, J.L. and Borg, S. (2000). Relationships between human-animal interactions and productivity of commercial dairy cows. Journal of Animal Science, 78 (11), 2821-2831.

Hendrickson, M. and Miele, M. (2009) Changes in agriculture and food production in NAE since 1945. In: McIntyre, B.D., Herren, H.R., Wakhungu, J. and Watson, R.T. (eds.) Agriculture at a crossroad, IAASTD North America and Europe, World Bank, Island Press, Washington, DC, USA, p. 20-79.

Hubbard, C., Bourlakis, M. and Garrod, G. (2007). Pig in the middle: farmers and the delivery of farm animal welfare standards. British Food Journal, 109(11), 919-931.

Hubbard, H. and Scott, K. (2011). Do farmers and scientists differ in their understanding and assessment of farm animal welfare? Animal Welfare, 20, 79-98.

IGD (2007). Consumer attitudes to animal welfare a report for freedom food by IGD available at http://www.rspca.org.uk/servlet/Satellite?blobcol=urlblob&blobheader=application%2Fpdf&blobkey=id&blobtable=RSPCABlob&blobwhere=1172248207766&ssbinary=true&Content-Type=application/pdf.

Ingenbleek, P.T.M. and Immink, V.M. (2011) Consumer decision-making for animal-friendly products: synthesis and implications. Animal Welfare, 20, 11-19.

Jensen, K.K., and Sorensen, J.T. (1998). The idea of 'ethical accounting' for a livestock farm. Journal of Agricultural and Environmental Ethics, 11, 85-100.

Jones, R.B. (1995). Habituation to human beings via visual contact in docile and flighty strains of domestic chicks. International Journal of Comparative Psychology, 8, 88-98.

Jones, R.B. (1997). Fear and distress. In: Appleby, M.C. and Hughes, B.O. (eds.) Animal welfare. CAB International, Wallingford, UK, pp. 75-87.

Jones, R.B. and Boissy, A. (2011). Fear and other negative emotions. In: Appleby, M. C., Mench, J.A., Olsson, I.A.S. and Hughes, B.O. (eds.) Animal welfare 2nd ed. CAB International, Wallingford, UK, pp. 78-97.

Jones, R.B., Blokhuis, H.J., De Jong, I.C., Keeling, L.J., McAdie, T.M. and Preisinger, R. (2004). Feather pecking in poultry: the application of science in a search for practical solutions. Animal Welfare, 13, S215-S219.

Kjærnes, U. (2012). Ethics and action: a relational perspective on consumer choice in the European politics of food. International Journal of Agricultural and Environmental Ethics, 25(2), 145-162.

Kjærnes, U. and Lavik, R. (2008) Opinions on animal welfare and food consumption in seven European countries. Welfare Quality® Report Series n. 2. Cardiff University, Cardiff, UK.

Kjærnes, U., Bock, B. and Miele, M. (2009). Improving farm animal welfare across Europe: current initiatives and venues for future strategies. In: Kjærnes, U., Bock, B.B., Higgin, M. and Roex, J. (eds.) Farm animal welfare within the supply chain: regulation, agriculture, and geography, Welfare Quality® Report Series n. 8, Cardiff University, Cardiff, UK, pp. 1-69.

Kjærnes, U., Bock, B., Roe, E. and Roex, J. (2008) Consumption, distribution and production of farm animal welfare. Welfare Quality Report Series n.7, Cardiff University, Cardiff, UK.

Kjærnes, U., Harvey, M. and Warde, A. (2007) Trust in food. A comparative and institutional analysis. MacMillian, London, UK.

Korte, S.M., Olivier, B. and Koolhaas, J.M. (2007). A new animal welfare concept based on allostasis. Physiology and Behaviour, 92, 422-428.

Lassen, J., Sandoe, P. and Forkmano, B. (2006).Happy pigs are dirty! – conflicting perspectives on animal welfare. Livestock Science, 103, 221-230.

LEI/CBS (1984). Landbouwcijfers 1984. Landbouw-Economisch Instituut/Centraal Bureau voor de Statistiek. 's Gravenhage/Voorburg, the Netherlands.

Lensink, B.J., Veissier, I. and Florand, L. (2001). The farmers' influence on calves' behaviour, health and production of a veal unit. Animal Science, 72, 105-116.

Lund, V. and Olsson, I.A.S. (2006). Animal agriculture: symbiosis, culture, or ethical conflict? Journal of Agricultural and Environmental Ethics, 19, 47-56.

Mayfield, L.E., Bennett, R.M., Tranter, R.B. and Wooldridge, M.J. (2007). Consumption of welfare-friendly food products in Great Britain, Italy and Sweden, and how it may be influenced by consumer attitudes to, and behaviour towards, animal welfare attributes. International Journal of the Sociology of Agriculture and Food, 15(3), 59-73.

McAdie, T., Keeling, L., Blokhuis, H. and Jones, R. (2005). Reduction in feather pecking and improvement of feather condition with the presentation of a string device to chickens, Applied Animal Behaviour Science, 93, 67-80.

Menghi, A. (2007). Italian pig producers' attitude toward animal welfare. British Food Journal, 109(11), 870-878.

Miele, M. (2011). The taste of happiness: free range chicken. Environment and Planning A, 43(9), 2070-2090.

Miele, M. and Bock, B.B. (2007). Competing discourses on farm animal welfare and agri-food restructuring. International Journal of Sociology of Agriculture and Food, 15(2), 1-7.

Miele, M. and Evans, A. (2010) When foods become animals, ruminations on ethics and responsibility in care-full spaces of consumption. Ethics, Policy and Environment,13(2), 171-190.

Miele, M. and Kjærnes, U. (2009) Investigating societal values on farm animal welfare: the example of Welfare Quality. In: Keeling, L. (ed.) An overview of the development of the Welfare Quality® project assessment system, Welfare Quality® Reports no.12, Cardiff University, Cardiff, UK, pp. 43-55.

Miele, M., Evans, A. and Higgin, M. (2009). Comparative citizen jury report. The results of a dialogue between citizens and experts regarding farm animal welfare in the UK, Norway and Italy. Welfare Quality® Reports no.18, Cardiff University, Cardiff, UK.

Miele, M., Veissier, I., Evans, A. and Botreau, R. (2011). Animal welfare: establishing a dialogue between science and society. Animal Welfare, 20, 103-117.

Millman, S.T., Duncan, I.J.H., Stauffacher, M. and Stookey, J.M. (2004). The impact of applied ethologists and the international society for applied ethology in improving animal welfare. Applied Animal Behaviour Science, 86, 299-311.

Moinard, C., Mendl, M., Nicol, C.J., and Green, L.E. (2003). A case control study of on-farm risk factors for tail biting in pigs. Applied Animal Behaviour Science, 81, 333-355.

Mormède, P., Andanson, S., Aupérin, B., Beerda, B., Guémené, G., Malmkvist, J., Manteca, X., Manteuffel, G., Prunet, P., Van Reenen, C.G., Richard, S. and Veissier, I. (2007). Exploration of the hypothalamic-pituitary-adrenal function as a tool to evaluate animal welfare. Physiology and Behavior, 92, 317-339.

OECD-FAO (2011) Agricultural outlook 2011-2020. OECD, Paris, France. Available at: http://www.oecd.org/site/oecd-faoagriculturaloutlook/48184304.pdf.

Paul, E.S., Harding, E.J. and Mendl, M. (2005). Measuring emotional processes in animals: the utility of a cognitive approach. Neuroscience and Biobehavioral Reviews, 29, 469-491.

Porcher, J. (2006). Well-being and suffering in livestock farming: living conditions at work for people and animals. Sociologie du Travail, 48, e56-e70.

Roe, E., Kjærnes, U., Bock, B.B., Higgin, M., Van Huik, M.M. and Cowan, C. (2009). Farm animal welfare in Hungary: a study of Hungarian producers, the meat retail market and of Hungarian consumers. In: Kjærnes, U., Bock, B.B., Higgin, M. and Roex, J. (eds.) Farm animal welfare within the supply chain: regulation, agriculture, and geography. Welfare Quality® Reports no. 8, Cardiff University, Cardiff, UK, 145-211.

Rushen, J., DePassille, A.M.B. and Munksgaard, L. (1999). Fear of people by cows and effects on milk yield, behavior, and heart rate at milking. Dairy Science, 82, 720-727.

Ruttan, V.W. (1998). The new growth theory and development economics: a survey. Journal of Development Studies 3 5, 1-26.

Skarstad, G., Terragni, L. and Torjusen, H. (2008). Animal welfare according to Norwegian consumers and producers: definitions and implications. International Journal of Sociology of Agriculture and Food, 15(2), 74-90.

Souquet, C., Kling-Eveillard, F. and Dockes, A.C. (2006). Les éleveurs de porcins parlent du bien-être animal dans les démarches qualité. Institut de l'Elevage, collection Résultats, Paris, France, 38 pp.

Steinfeld, H., Gerber, P., Wassenaar, T., Castel, V., Rosales, M. and De Haan, C. (2006) Livestock's long shadow: environmental issues and options. FAO, Rome, Italy.

Stott, A.W., Milne, C.E., Goddard, P. and Waterhouse, A. (2005). Projected effect of alternative management strategies on profit and animal welfare in extensive sheep production systems in Great Britain. Livestock Production Science, 97, 161-171.

Terragni, L. and Torjusen, H. (2007). Norway. In: Evans, A. and Miele, M. (eds.) Consumers' views about farm animal welfare. Part I: national reports based on focus group research. Welfare Quality® Reports no.4, Cardiff University, Cardiff, UK, pp. 253-322.

Tovey, H. (2003). Theorising nature and society in sociology: the invisibility of animals. Sociologia Ruralis, 43(3), 196-213.

Turner, S.P. and Dwyer, C.M. (2007). Welfare assessment in extensive animal production systems: challenges and opportunities. Animal Welfare, 16, 189-192.

Verbeke, W. (2009). Stakeholder, citizen and consumer interests in farm animal welfare. Animal Welfare, 18(4), 325-333.

Vos, J.A.M. (1997). Schrap individualism uit uw woordenlijst. Pluimveehouderij, 27, 10-11.

Waiblinger, S., Boivin, X., Pedersen, V., Tosi, M., Janczak, A.M., Visser, E.K. and Jones, R.B. (2006). Assessing the human-animal relationship in farmed species: a critical review. Applied Animal Behaviour Science, 101, 185-242.

Winter, M., Fry, C. and Carruthers, S.P. (1998). European agricultural policy and farm animal welfare. Food Policy, 23, 305-323.

Chapter 3. Animal welfare: from production to consumption

Henry Buller

3.1 Introduction

As the previous chapter in this volume has described, the last decade has seen a significant rise in societal concern for farm animal welfare in many European states and beyond. We might say of farm animal welfare that it has become almost 'mainstreamed' into a broader general awareness, into policy and regulatory rhetoric and, to a large degree, into livestock farming practice.

For many commentators, the starting point of contemporary farm animal welfare policy and the inauguration of animal welfare science was the publication in the United Kingdom of Ruth Harrison' book '*Animal machines*' in 1964, followed by the UK Government's Brambell Committee Report (1965). Both of these publications were highly critical of the poor welfare practices that were in many ways intrinsic to 'factory farming' methods. However, if Harrison's book and the Brambell Committee's report were of vital importance in prompting the subsequent regulatory and institutional framework for farm animal welfare, they were also equally important in turning it into an issue for contemporary society across many different countries. Rachel Carson, the American biologist, writing in the preface to Harrison's book made the following appeal:

> I hope it will spark a consumers' revolt of such proportions that this vast new agricultural industry will be forced to mend its ways.

That 'consumers' revolt' has, we might argue, only ever been a partial phenomenon, and hasn't taken place in equal proportions everywhere within Europe and beyond. Yet, the opening up of modern industrial husbandry practices to scrutiny that these publications ushered in, and the ensuing spread of concern across many European states, has indubitably helped transform farm animal welfare into an issue that is increasingly regulated from, and driven by, the 'demand' end of the food chain, whether by the consumers and citizens themselves or by the retailers and food services that directly supply them. That is not to say that farm animal welfare is any less a production concern, or any less one that engages the commitment of the majority of those caring directly for farm animals. What it does mean however, is that, as a societal issue, farm animal welfare has become increasingly exposed to different forms of demand-led governance. Fundamental to both has been consumer choice and the role of the market.

Henry Buller

Farm animal welfare is today a significant element of segmentation within a food marketplace that is increasingly differentiated in response to greater consumer demand for choice, regardless of whether that choice is driven by issues of price, quality or ethical commitment. Claims of higher levels of farm animal welfare are now mobilised both as an additional component of quality (Buller and Roe, 2012), thereby contributing to the higher prices being demanded of certain products within retail outlets, and as an element of increasingly selective access to retailers' shelves based upon the meeting of welfare criteria. Farm animal welfare standards, along with other considerations, are now regularly used to differentiate and even rank the ethical commitments of different food retailers. Some farm animal welfare NGOs run welfare award schemes for supermarket chains. Moreover, farm animal welfare is emerging as a distinct retail niche in its own account, with distinctive labelling schemes, such as, perhaps most notably, the UK RSPCA's *'Freedom Food'* scheme, now existing in a number of European countries (e.g. the German *'Neuland'* scheme or the French *Thierry Schweitzer* product label). In becoming an issue of consumption, farm animal welfare is today an important element in retailer (and therefore market) strategies with attention increasingly being given to forms of consumer communication, packaging and labelling: how to get the message across, how to improve consumer knowledge and how to facilitate consumer choice in favour of products that are derived from systems with socially acceptable or higher welfare practices.

Thus, farm animal welfare has become the latest rung of the ethical ladder both for individual food consumers and for nations and international institutions for whom the oft-quoted words of Gandhi remain ever pertinent ('The greatness of a nation and its moral progress can be judged by the way its animals are treated'). That this growing concern should emerge at a time when more animals are killed for human consumption than ever before is perhaps a source of both comfort and astonishment. The sheer industrialisation of contemporary global livestock farming and slaughter generates its own plethora of welfare (as well as many other) issues over and above more traditional ethical concerns about such issues as the 'right' to kill and the duty of care. The new information technologies offer new possibilities for understanding how animal products come to our plates thereby enabling consumers to make more informed choices. Ethics is the new market place and farm animal welfare one of a number of new ethical parameters that are employed to create markets (Webster, 2001). Whether this new societal interest has improved farm animal welfare is a more difficult question to answer. It has raised awareness, it has created markets for more welfare-friendly production systems, it has helped promote the release of national and international funding for research in animal welfare science but it has also exposed the scale of the task at hand and the hugely variable contexts in which animals are raised, killed and consumed across the Globe. At this time, farm animal welfare cannot be seen in isolation from issues of human welfare, of environmental change and so on.

Perhaps the most significant shift then, of the last decade, is that it has become a human as well as an animal issue, an issue for all consumers, not merely an issue for producers keen to produce healthy and therefore profitable animals. It is an issue for politics as well as for science.

With this in mind, this chapter considers animal welfare as a contemporary societal demand and describes some of the work carried out in the Welfare Quality® project and elsewhere that has helped to illuminate the role of two critical sets of social agents (i.e. consumers/citizens and retailers) in both driving and structuring new forms and intentions of farm animal welfare governance and practice.

3.2 Citizens as consumers of welfare

Consumers, for a wide variety of often complex reasons are increasingly concerned about the welfare of farm animals (Roex and Miele, 2005). In response, food chain actors, from retailers to processors, recognise that higher welfare can be an added value component of product quality and therefore a mechanism for either creating price gains, trade advantage or brand difference or for assuring customer fidelity. A third group, the producers, more and more aware of both consumer concerns and retailer strategies, and yet mindful of their own responsibility and professional status as stockpersons are increasingly responding to the need to maintain and improve welfare standards through their incorporation into good practice.

For much of its albeit recent history, farm animal welfare has been governed by an expanding raft of regulation which includes not just formal laws and rules but also codes of practice, certification procedures and mandatory training in certain areas of animal care, transport and slaughter. With the emergence of consumer choice and ethical purchasing, however, a new mechanism of governance appears; that of the market.

Making it easier for informed consumers to select products coming from higher welfare systems, or to select brands or even retailers for which farm animal welfare is an explicit criterion has, over the last 15 or so years, become a highly significant and competitive driver of welfare improvements. While public regulation and official codes of practice tend to fix welfare knowledge and standards at a given point in time, competitive marketing allows food chain actors to drive private standards upwards, as long as the market demand sustains this. Increasingly therefore, we see developing in many countries a dual regulatory system. On the one hand, laws, statutes and codes provide a base-line mandatory level of welfare standards to which all must comply. On the other hand, private and industry standards, certifications and labels respond more rapidly to perceived and actual consumer and citizen demand, NGO pressure, scientific research and strategic market positioning.

The Welfare Quality® research project has clearly shown that a significant number of consumers describe themselves as concerned about the welfare of farm animals and feel that steps ought to be taken to improve the animals' quality of life across different production systems. The well-known and oft-quoted Eurobarometer studies of 2005 and 2007 (Eurobarometer, 2005, 2007) suggested a similar concern while other consumer studies have subsequently confirmed it, often offering analysis of how that concern is both constructed and articulated.

There are two critical interrogations here which form the principal arguments of this chapter. The first, which has been the central preoccupation of the Welfare Quality® research, concerns the availability, the accuracy, the veracity and the communicability of welfare information within the food chain, but particularly that which is accessible to the consumer. How are good welfare practices to be assessed and communicated as they transit the different nodes of the food supply chain? How are products coming from animals enjoying higher welfare standards to be identified by those seeking to buy such products or by those simply concerned by farm animal welfare as an issue? I shall return to these fundamental questions later in the latter half of this chapter.

The second critical point, for those actively engaged in promoting higher welfare systems, is the apparent 'gap' between attitudes on the one hand and action or behaviour on the other (IGD, 2007; Kjaernes 2005; Kjorstad, 2005; Vermier and Verbeke, 2004). Concerned consumers do not necessarily purchase products from higher welfare systems even if these are identifiable as such. Where they are not identifiable, consumer concern remains hidden. Attention has thus become focused on the means of effectively translating stated concern into active purchasing behaviour. Such attention however pre-supposes that consumers want to be able to act on this issue and that they feel that this is something for which they should be responsible. This should not always be so assumed.

As a start to understanding the attitudes/action gap, one approach employed in the Welfare Quality® research has been to break down that rather cumbersome notion of 'the consumer' into a number of different sub-groups. For the purposes of simplicity, four are suggested here.

3.2.1 Consumers for whom welfare is a central concern

First, an acknowledged number of consumers do tend to actively seek out and buy products from production systems that explicitly adopt welfare parameters that are significantly higher than legal requirements. Although this is possibly a relatively small sub-group of consumers, it raises three important questions: first, whether or not the demand from these consumers is being met by higher welfare production systems, second, whether correct and verifiable information about those systems is being accurately communicated to these consumers (see below) and, finally, whether this consumer sector has the genuine potential to substantially expand. It has been recognised by many food chain actors that already this particular group of consumers could potentially generate significant market opportunities for higher welfare production systems. The Institute of Grocery Distribution (IGD) survey for Freedom Food (2007) estimated such regular consumers as representing around 10% of the market (with an additional 34% buying 'some' welfare products). The study by Brooklyndhurst for the UK's Department of Environment, Food and Rural Affairs (Defra), published in 2010, set the number at 30% (Brooklyndhurst, 2010). However, as the IGD conclude, this is still too few to have a sustained impact. To achieve a durable impact upon production systems, the proportion of consumers making food choices on the basis (at least in part) of welfare needs to increase. An important caveat we might introduce here is that, as is often the case, free range eggs and Organic systems are often uncritically taken as synonymous with 'higher welfare'. Indeed, in many of the aforementioned studies the results on higher welfare purchasing are almost all accounted for through purchase of eggs coming from free-range systems.

3.2.2 Consumers for whom welfare is a component of 'quality'

A second group of consumers like to think they are purchasing higher welfare products through what we might call 'quality product' chains. Elsewhere, we have referred to this as the '*terroir*' model (Kjaernes *et al.*, 2009), where an appreciation of the animal is bound up with notions of taste, tradition and the socio-territorial context of production. The issues here are firstly, do such 'quality' chains provide genuine and confirmable welfare gains that can be effectively assessed, and, secondly, how important is welfare to these consumers as a distinct and necessary component of these quality ranges. As the Welfare Quality® research demonstrated, this is a far larger group of consumers than the first sub-group, proportionally higher in some countries than others. These consumers make the implicit link between higher welfare standards on the one hand and quality production systems bearing some form of quality label (*Appellation d'Origine Contrôlée* (AOC), *Label Rouge* and so on) or even distinct national provenance on the other, even when this link is neither explicit nor is necessarily based on certifiable standards. In some countries this represents a significant sector of food buyers. In France and Italy, for example, quality labels are often assumed to contain higher welfare conditions (Buller and Cesar, 2007) while in Norway there is a strong popular belief that home-produced food meets higher welfare standards than imported food – a belief shared by many UK consumers (Kjaernes and

Lavik, 2008). However, this is a potentially difficult group to target through specific appeals to welfare-oriented purchasing, first, as the welfare benefits and claims of such 'quality' systems would have to be accurately assessed and confirmed through agreed conventions and, second, specific welfare claims and benefits would have to be distilled out from other 'quality' claims. Evidence from Welfare Quality®, suggests that buyers of such products would not be prepared to give up on what they perceive as a broad sense of quality for the specific sake of welfare. In this context, welfare becomes a 'bundled' element in perceived product quality and not a 'stand-alone' component of product quality.

3.2.3 Hidden consumers

A third group of consumers, perhaps the majority group in many countries, do not actively buy or even seek higher welfare products but maintain that they would prefer to do so, all other things being equal. What emerges clearly from the Welfare Quality® focus group research (Evans and Miele, 2007, 2008) is that a significant proportion of consumers would like to be able to make more food choices based upon the welfare of farm animals if they could. The issue here is the nature of the barriers inhibiting a more active engagement in the purchase of products from higher welfare systems. Most commonly, such barriers are framed either in terms of the equal and sometimes greater importance of other concerns, be they price, access and so on or the lack of suitable and appropriate information enabling the selection of higher welfare products to be made. It is notable that this group has become a key 'target group' of contemporary welfare labelling, retailer branding and choice editing strategies.

Of course, one major barrier is price; identified by a number of recent studies (Brooklyndhurst, 2010; Evans and Miele, 2007, 2008; University of Hertfordshire, 2010) as the most important obstacle to potential consumers. Yet, higher welfare products are higher priced, not necessarily because they are more costly to produce, but also because they constitute an important and distinctive component of market specialisation. Many higher welfare products are often sold in premium ranges, commanding higher prices largely because these can be obtained from discerning and committed consumers (those of the first group identified above). One might argue that not only are higher production costs (where they occur) being passed on directly to consumers but that additional revenues are being generated on such products because of their premium status.

Additionally, price rarely acts alone as a deterrent:
> A closer examination of the focus group discussions indicates that there are several practice/sociological factors, which could intervene to rule out any simplistic interpretation of the link between price and desire to purchase. Indeed, it would seem

that many participants were actually far more interested in the balance between quality and price than in price alone (Evans and Miele, 2008: 86).

Leaving aside price for the moment, food choices based upon welfare criteria are also not being made because of a lack or a confusion of claims and messages being presented to consumers at point of sale. Moreover, this confusion can be compounded by a lack of detailed knowledge or understanding of contemporary food animal production systems. This possibly leaves consumers susceptible to overly-simplistic (and sometimes erroneous) associations in their assumptions about welfare and provenance. Evans and Miele again (2008: 42):

Participants' knowledge about farms was heavily influenced and framed by the notion that one can make an almost bi-polar distinction between highly intensive 'factory' farms (which are considered to have extremely low welfare) and alternative farming systems, such as organic, free-range, outdoor access, traditional or small scale systems (which are considered to have higher levels of welfare).

Yet, as these authors go on to demonstrate, this common notion contrasts strongly with a great deal of contemporary welfare science and evidence. Finally, as also widely reported in the Welfare Quality® research, there are issues of consumer mistrust over what information is provided by retailers and food companies regarding the role and place of welfare considerations within existing production systems. For the European Commission, this can impact directly upon purchasing choice:

Unless they have reliable knowledge about the added-value of animal welfare-friendly products, they cannot be expected to pay a price that reflects the higher product quality (Commission for the European Communities, 2009: 6).

The issue of information and welfare labelling will be tackled in the following section of this chapter.

3.2.4 Consumers and ethical proxies

It is often assumed by those promoting higher welfare standards that consumers wish for farm animal welfare to be an area of shopping 'choice'. While many of the studies cited above clearly demonstrate a sense of consumer concern over farm animal welfare, it is by far from certain that the majority of consumers want to have to make the choice of buying animal products derived from production systems whose welfare standards go significantly higher than legal minima over those from production systems which merely conform to legislative requirements. Hence, a final group of consumers simply do not want the responsibility of making the choice themselves but nonetheless believe the choice should be made. Critically, the issue is who should make these choices and where in the supply chain should these choices be made.

These have been central questions in one of the main sub-projects of the Welfare Quality® project.

This fourth group of consumers and food buyers largely place responsibility on the retailer (or on other food chain actors) rather than themselves to act ethically. They feel that farm animal welfare is not their responsibility but rather that of those engaged in ensuring that such systems conform to socially accepted standards (Frewer *et al.*, 2006). The Eurobarometer research in the mid-2000s showed that consumers would largely prefer that welfare was to a degree 'taken care of' before products get to the supermarket shelf. This is important because it has implications for the nature of the governance of welfare that might, at the end of the day, be adopted, or indeed for the relevance of such governance at all.

3.3 The retailer response: creating and managing consumer choice

Today, in many, particularly European, countries large retail companies constitute the undoubted 'captains' of the food sector. They are no longer simply the 'end-users' and mere 'points of sale' for extended supply chains dominated by marketing boards, manufacturers and producer groups. Undoubtedly, one of the more marked characteristics of the last 30 or so years has been not only the dramatic growth and concentration of food retailers (Burch and Lawrence, 2007; Hughes and Reimer, 2004) but also the extension of their power and controlling influence all the way down the food supply chain (Busch, 2010; Marsden *et al.*, 1998). Through their considerable purchasing power, retailers are increasingly able to dictate terms and conditions to producers and suppliers (Grabosky, 1994; Hughes *et al.*, 2010) thereby obtaining more and more favourable terms for themselves. Moreover, they are becoming increasingly adept at creating markets and structuring the very consumer demand to which they claim to respond both through complex processes of product segmentation, thereby facilitating greater consumer choice, and yet also through the effective editing of that consumer choice.

3.3.1 Assurance

The vertical integration of food supply chains and the growing dominance of the retailers was initially driven by the need to achieve maximum supply chain efficiency (thereby enabling them to compete on lower prices to consumers), to reduce risk and to assure product standardisation without actually investing in production itself, with all its associated costs and hazards (Godley and Williams, 2007). However, new concerns for the maintenance of product and brand quality, and hence the continuing loyalty of their client consumers, as well as conformity to growing legal requirements regarding food safety, has meant that retailers have progressively engaged in extending standard setting throughout their supply chains across three critical domains; product quality, product safety and, more recently areas of identifiable consumer (and often NGO) concern, such as farm animal welfare (Hobbs, 1996) and environmental sustainability (Fulponi, 2006; Lockie *et al.*, 2006). This has also brought with it, in addition to increasingly tightly defined production contracts, new forms of supply chain management; the tri-partite regime where retailers, in Busch's words:

> Demand that their suppliers adhere to a set of stringent standards. To ensure that the suppliers adhere to the standards; to avoid the costs of checking; and so as not to cast themselves in the role of police officers, the lead firms require that some third party certify that the suppliers are operating in conformity with the standards. Furthermore, in order to ensure that the certifiers themselves behave as expected, the lead firms demand that certifiers be accredited by international accreditation bodies (Busch, 2010: 7).

Recent years have seen the rapid growth and development of private forms of food chain regulation such as quality assurance schemes, private standards operated by major food retailers, NGOs and others, third party certification and formal accreditation. Retailers have been particularly vigorous in using these new non-governmental regulatory devices, particularly to demonstrate compliance; the compliance of their suppliers with existing mandatory production standards and retailer compliance with the requirement, under national and EU law (notably EU Regulation 178/2002) to show that 'due diligence' has been taken to ensure customer safety and satisfaction. In part also, they are used to maintain and protect brand identity, particularly for retailer own-label products, and customer loyalty. Finally, such private forms of regulation underscore intentional commercial strategies to entice discerning consumers to part with greater amounts of money (Murdoch, 2005: 112). Consequently, these private food standards, that generally either confirm or exceed mandatory legal minima, have become extremely commonplace within the food sector (Henson and Reardon, 2005) and are a growing feature of the animal welfare commercial landscape. A 'retail audit' carried out under the Welfare Quality® project (Roe and Marsden, 2007) identified a significant range of food products containing specific welfare claims available in European retail outlets. Although there is considerable variation across European states, the growing role of retailers in promoting such claims, and backing them up with their own assurance procedures, is evident. This led Murdoch to observe that: 'clearly, the major retailers are key players in the formulation and implementation of any strategies that aim to raise welfare standards' (Murdoch, 2005: 112).

Thus, farm animal welfare has emerged as one of the key areas for quality assurance and private standards (FAWC, 2001; Hobbs, 1996) having already reached what the UK Farm Animal Welfare Council (FAWC) describes as a 'critical mass' by the beginning of the current century (FAWC, 2001: 30). Today, animal welfare components of certification, auditing and assurance schemes are operated by a wide range of non-governmental actors, including the animal production industry, food manufacturers, trade associations, food services, NGOs, food retailers (Buller and Roe, 2010; Farm Sanctuary, 2009; FAWC, 2001; Compassion in World Farming (CIWF), 2002; Roe and Higgin, 2008) and even, some have recently argued, veterinarians (Fordyce, 2011). In some countries, notably the United Kingdom, they cover, in one form or another, the vast bulk of farm animal production output in the majority of key sectors.

Assurance, auditing and certification schemes operate alongside legally defined minimum welfare standards. Some act to merely confirm that such minimum legal standards have been met, particularly those that link such assurance with national or sub-national provenance. Others, however, require standards that exceed these legal minima, in some cases, significantly. FAWC's 2001 review maintained that, of the UK supermarket schemes operating at the time, at least half went beyond the legal requirements of legislation and codes of practice.

The dramatic growth of quality assurance, particularly on the part of food retailers, and often pushed by a vibrant and supportive NGO sector (Buller and Roe, 2010) has unquestionably raised the profile of farm animal welfare within the food purchasing and consumption environment in what are often innovative ways that explicitly re-connect consumers not only with producers but also with their livestock. Additionally, this competitive assurance environment has led to a far faster improvement of many welfare criteria than would have been achieved by the more cumbersome process of legislation or the more static approach of uniform standards. Balsevich *et al.* (2003: 1153), in another context, have argued that 'markets segmented by quality and price may serve the varied needs of consumers better than a single minimum quality standard'. Because most supermarkets will now only buy food from suppliers that farm to their specified standards, the development and spread of such standard regimes has multiplied to become virtually inclusive. Indeed, suppliers lying outside those regimes are finding it increasingly difficult to sell their product within conventional supply chains even if they are producing to levels of regulatory compliance (Henson and Reardon, 2005).

With this expanding universality of assurance schemes has come, perhaps paradoxically, their declining visibility to consumers. While farm animal welfare has certainly gained prominence on retailer shelves, and higher proportions of animal-based products now derive from assured production chains, the Welfare Quality® research into retail practice reveals that scheme standards themselves are not always made explicitly accessible to consumers either at point of sale or on retailer web sites. Comparing the UK experience to other Member States, Roe and Marsden (2007: 66) write:

The moves, particularly by UK retailers to reduce the use of logos on their own brand products, or to use them as a market segmentation strategy pushes the assurance schemes back towards a predominantly industry concern. This leads to large amounts of meat and dairy products that are produced to higher welfare levels than EU minimum standards but which are not labelled as such. However, where the retailers are less dominant, the place of the label is still thriving on food packaging for products produced by manufacturers or farming cooperatives.

It is important to remember that animal welfare has become a significant new area of product segmentation and differentiation as well as brand competition reflecting, in Henson and Reardon's (2005: 252) words 'the growing predominance of quality as the mode of competition in agri-food systems'. As one French supermarket purchasing manager pointed out during the course of the Welfare Quality® research (Buller and Cesar, 2006):

> Eggs is a very banal sector, a simple undifferentiated basic product that we have specifically sought to segment. With a generally more wealthy clientele, animal welfare can be an interesting idea. It is all marketing though, for the egg itself, there is absolutely no difference; for the hen yes, they are less stressed outdoors.

Animal welfare is a marketing device. Not only do retailers compete with each other on quality (both in 'ethical' and in health/taste terms) but they also seek to segment their product ranges to cater for different client groups, as the above quotation reveals. Growing reference to farm animal welfare in marketing strategies should not be seen as a straightforward response to consumer demand but rather as an element in strategic marketing, brand positioning and the 'editing' of consumer choice. On the one hand, the inclusion of differential welfare claims (for example, in broiler stocking densities and growth rates (RSPCA, 2006)) across different product ranges offered by most large food retailers provides a carefully constructed response to structured consumer choice and the potential for higher profits in some areas. Some retailers may themselves offer four or five different sets of welfare standards, starting with those assurance schemes based upon legal conformity and rising to organic certification and schemes such as the RSPCA's Freedom Food. Yet, on the other hand, the virtual disappearance of caged-eggs from many large chains across Europe over the last couple of years, provides clear evidence of intentional choice editing, where consumers are no longer offered the choice to purchase a product (in this case fresh eggs from caged systems) that the retailer feels is both of an unacceptable standard of welfare and detrimental to the image the retailer wishes to promote of itself as an ethical establishment (Buller and Roe, 2012).

There have been clear benefits from the growing incorporation of farm animal welfare into food marketing strategies. Yet there are also a number of additional considerations that merit exploration (Buller, 2010; Buller and Roe, 2010). First, there is the issue of the complex relationship between market-driven assurance and mandatory regulatory standards. To be effective as a basis for competitive differentiation, private schemes need to be, to a greater or lesser extent, exclusive. If all actors supplied to the same level, there would be no competitive advantage. The viability of a higher welfare scheme will therefore depend upon the continued existence of lower welfare production lines and, unless it is very successful (as has been the apparent case of free-range eggs), it may not have a dramatic impact upon production systems as a whole. It is for this latter reason that the development of effective market segmentation strategies, and the innovative animal welfare research that often accompanies them (collectively contributing to a raising of standards across the board) should also feed into the regular review of legislative minima and official codes of practice.

Second, there is the issue of communication; whether the right sort of information is available and is appropriately communicated to enable consumers to make informed decisions about buying products from higher welfare systems. Recent years have seen a number of organisations, backed up by research, claim that the information available to consumers does not allow them to make choices based upon the welfare status of the animals involved (for example FAWC, 2011). The common claims are that the information available is either non-existent, is insufficient, is misleading or is simply not presented in a comparable form, thereby making it difficult for consumers to make informed choices. There is, it is often claimed, no standardised system of information. The information on the welfare conditions of a chicken sold from one retailer's 'Quality' ranges might be in a substantially different form to that from another's. The issue here then is not just one of information availability and information clarity but also of information standardisation.

3.3.2 Labelling

The labelling, at point of sale, of food products coming from animals enjoying higher standards of welfare has become, in recent years, the focus of considerable political debate within Europe and beyond (Commission of the European Communities, 2009). It has been fed, in part, by the apparent popularity of existing production system labels associated with organic production, which offer a *cas de figure* for an eventual 'welfare label'. It has also been driven by the numerous pieces of research undertaken over the last 10 years which conclude that consumers themselves are not only confused by the multiplicity of existing welfare claims and labelling practices but are also frustrated in their attempts to engage with animal welfare as a result. The evidence from these surveys suggests that if there was more, clear and accessible information available, consumers would buy more products of a higher welfare provenance. Yet, Miele (2010:

4), drawing on the international comparative Welfare Quality® research (Evans and Miele, 2007, 2008), qualifies this widely accepted assumption on the following basis:

> Reaction to more information on this issue would lead to different courses of action: some would protest for more/better regulation on this matter (mostly Scandinavian countries), others would seek to change to more welfare-friendly food retailers (preferred option in Italy and the United Kingdom), whereas other consumers would both *boycott* the products perceived as cruel to animals and, possibly, if price, availability and other circumstances would allow them, would choose animal friendly products.

Evidence would suggest that the preferred form of that information – from the consumer's point of view – is some type of label or logo. The Eurobarometer studies showed a preference amongst consumers for some form of welfare labelling and, in recent years, a number of studies have explored the potential for such labelling. The European Commission's own assessment reveals:

> Both the analysis of the outcome of two Eurobarometer surveys and the feasibility study on animal welfare labelling suggest that animal welfare labelling, based on sound scientific knowledge and assessed on the basis of harmonised requirements, could enable consumers to make informed purchasing decisions and make it possible for producers to benefit from market opportunities (European Commission, 2009).

It is worth noting that a welfare label is seen not just as a means of identifying higher welfare products and harmonising or standardising welfare claims but also, even for those committed existing consumers of higher welfare products, as a way of lowering the search costs for higher welfare goods (Vanhonacker and Verbeke, 2009; Vanhonacker *et al.*, 2007). Interestingly, and in contradistinction to many of the assumptions usually made about welfare labelling, Vanhonacker *et al.* (2007) see a welfare label as being of greater value to consumers who already make decisions based upon welfare than on those who do not (the latter being consistently put off by the higher prices of labelled products).

Drawing on the Welfare Quality® research, Mayfield *et al.* (2007: 71) suggest:

> Consumers are generally in favour of welfare product labelling with an assurance scheme to signify the animal welfare provenance of meat and other animal products. Swedish consumers do not feel that this was as important as their GB and Italian counterparts, probably because they appeared to have more trust in their own farming systems. A significant proportion of consumers is also in favour of a welfare grading scheme. Most consumers had a positive willingness to pay for higher welfare friendly food products.

Absence of suitable and harmonised information leads then to significant calls for some form of welfare labelling. But this is just the beginning. The critical issues are, first, what kind of labelling, second, what is actually being labelled and, third, how might labels relate to other existing forms of market segmentation.

Looking at different examples of segmentation, we can identify a number of approaches:

- *Explicit welfare labelling*: where a product is labelled according to the welfare criteria applied to the system of production over and above regulatory minima.
- *Quality assurance labelling*: where a product is labelled as conforming to an existing assurance or quality scheme for which higher welfare standards are claimed.
- *Conformity labelling*: where a product is labelled as having certain characteristics that might be assumed to be in conformity with regulatory welfare requirements.
- *Omni-labelling*: where a range of ethical considerations are brought together into a composite label or scheme.
- *Production system labelling*: a product is labelled according to the type of production system and according to an established and defined range of systems.
- *Criteria labelling*: where a label might refer to a single criterion that is either specifically focused on welfare or assumes some welfare advantage.
- *Branding*: where a product is labelled as being part of a brand to which certain welfare conditions are specified (but are not necessarily presented on the packet itself).

Many examples of these different approaches exist across Europe (for a review from the Welfare Quality® research, see Roe and Marsden, 2007). For the individual consumer, however, the proliferation of schemes is a source of considerable confusion:

> Firstly, over consumer information on the comparative welfare benefits of one process or procedure or set of claims over another, secondly over the extent to which these various claims go further than legal regulatory compliance and thirdly, over the precise nature, and welfare implications, of the terms employed, such as 'free range' or 'outdoor' (FAWC, 2011)

A final implication of the growth of private welfare standards, particularly at the consumer-facing end of food supply chains is their potential impact upon understandings of farm animal welfare itself. In simple terms, certain aspects of farm animal welfare lend themselves more readily to product marketing than others (Buller and Roe, 2012). The difference between popular citizen/consumer notions of farm animal welfare on the one hand and scientific and farmer notions on the other has been well reported in the Welfare Quality® research (Evans and Miele, 2007, 2008; Kjaernes and Lavik, 2008) and elsewhere (Pricket *et al.*, 2010; Vanhonacker *et al.*, 2007). Emphasis on the consumer-friendly dimensions of farm animal welfare within private assurance schemes, such as access to 'outdoors', grass feed, 'natural behaviour' and so on, may not only lead to the obfuscation of the welfare issues inherent in these dimensions (such as, for example, the higher mortality rates of 'free-range' systems) but also lead to a neglect of those other welfare aspects that, from a welfare science point of view, may be considerably more problematic (such as lameness, pre-slaughter stunning, transport).

3.4 Ways forward

First and foremost, animal welfare can and should remain an issue where market segmentation through private labelling and assurance schemes is allowed to continue. There are two reasons. First, through the work of NGOs (who to some extent speak not only for the animals but also for the consumers), through the work of those retail actors either ethically committed to improving welfare or recognising the possibilities of a market segmented by differential welfare claims and through the work of informed producers who seek to benchmark and sell their own products, many aspects of farm animal welfare have improved dramatically in the last 10 or so years – both in real terms and in terms of public awareness. Second, because if we accept that more consumers will purchase products from higher welfare systems if these are identifiable as such, then these products ought to be differentiated in a manner that responds to this demand.

Where it gets problematic, particularly for the consumer, is in the different ways in which farm animal welfare is assessed. Elsewhere (Buller, 2009; Buller and Roe, 2008; Roe *et al.*, 2011) we have called for a commonly agreed set of parameters for welfare assessment, a position endorsed by the Commission in its consultation document of 2009.

> Certain stakeholders supported the principle that the more private animal welfare labels are present on the retailing market, the more we need to determine a methodology to assess and compare animal welfare standards (Commission, 2009: 5).

A consistent call from the Welfare Quality® project has been the need for: (1) a reliable, robust and animal-based welfare assessment system; (2) far greater consistency/harmonisation of welfare assessment measures and mechanisms that can be employed in these various standards; and (3) the more coherent and comparable presentation of welfare information to consumers (see Chapters 1, 3, 5 and 8).

A second way forward, one that has moved in and out of fashion over the last few years, has been that of a 'higher tier' welfare label, similar to the current UK Freedom Food scheme (or near parallels in other countries). The British Veterinary Association (BVA) has stated its clear wish to see: 'the development of a clear welfare label that consumers recognise as a mark of higher animal welfare', while FAWC, in its 2006 report, also argued that 'the Government should press at EU level for a single, accredited, mandatory labelling system on animal welfare grounds to be agreed by stakeholders and used for all animal-based products' (FAWC, 2006: 26). The Commission, in its own assessment of the potential for an EU welfare label, while acknowledging the place of private and other market-driven labelling schemes, nonetheless explores the potential of such a broad 'higher tier' scheme:

> Some existing labelling schemes address animal welfare requirements alongside other quality standards, such as organic farming or environmental protection. For this reason, the added value of a possible EU animal welfare labelling scheme in view of existing schemes will need to be carefully considered, as well as how to avoid any possible overlap (Commission, 2009).

Hence, the RSPCA's 'Freedom Food', along with other schemes in other countries, run by NGOs, retailers and other actors, would continue to occupy a competitive segmented place within the marketplace.

Despite the launch of the Commission's debate on the issue of welfare labelling in 2009 (Commission of the European Communities, 2009; GHK, 2010; Horgan and Gavinelli, 2006), the Council's recent agreement on the provision of food information to consumers of December 2010, makes no specific reference to welfare labelling and rejects the more recent proposal of the Parliament to label meat from animals slaughtered without stunning.

Yet, most recently, research commissioned by the UK DEFRA concludes:

> The evidence presented in this research suggests that animal welfare labelling on its own would have a limited effect on purchasing behaviours: informational barriers are not currently the main reason why most UK consumers do not translate their values and attitudes around animal welfare into action in the supermarket (Brooklyndhurst 2010: 73).

While the Commission itself, in its consultation document of 2009, betrays a certain caution:

> Clearly any such EU scheme, which could avoid segmentation of the internal market as well as facilitating intra-Community trade, would need to demonstrate that it can add value both to existing private schemes and to the organic regulation without harming them (Commission 2009: 4).

The debate here is very much on-going and includes a large number of considerations: whether a universal voluntary 'higher scheme' would be attractive to food suppliers, how it might relate to existing 'higher' welfare level schemes, how it should articulate with the debate over method of slaughter and so on. For animal welfare is rarely, if ever, considered a 'stand-alone' issue for the retail sector. It is always bundled together within a wider set of consumer concerns that combine, to a greater or lesser extent, product quality, taste, animal health, environmental sustainability, a countryside aesthetic and an awareness of animal lives. Consider the responses of two retailer managers to part of the Welfare Quality® research; one from a major French retail chain, the other from a major British retailer, both companies well known for their insistence on food quality, provenance and welfare:

> Yes and that is all about eating quality because the other thing to say is you can have all this fancy animal welfare, but if it doesn't taste good there is no point in having it so it has to be the whole package, animal welfare is a contributory factor to good eating meat (Manager, UK store – from Roe and Higgin, 2008).

> I cannot imagine a label that will be uniquely about welfare because there is no difference in the final product and, in the current climate, it has to be associated with an awareness of the environment, a more global concern than that for animal welfare (Quality Manager, French Supermarket Chain – from Buller and Cesar, 2006).

Additional complexity is introduced by the very different nature of farming systems, from intensive poultry units to free range beef production, and the difficulty of establishing a common lexicon of welfare terms to cover all. Moreover, as the Welfare Quality® research has so amply demonstrated, there are significant social and cultural differences in European attitudes towards welfare that would complicate the wider adoption of a single European label.

A third way forward and one that has recently been endorsed by the Farm Animal Welfare Forum (FAWF, 2010) is that of production system labelling. At first glance, this approach might have some advantages. For example, it is more generalisable – or at least more easily understood ¥ across different social or cultural contexts than a welfare-based label. Moreover, it is an inclusive system, rather than an exclusive one, allowing all production systems to be appropriately identified. Finally, it is flexible, permitting the use of new parameters to be brought into consideration.

However, there are also some significant limitations to such an approach. First, it relies on a robust and transferable definition of different production systems; this is a difficult thing to accomplish. Second, there is a real danger that it is prone to over-simplification. This is certainly a concern with the growing – and often deceptive – use of 'indoor/outdoor' criteria as de facto welfare indicators. Third, it implies a return to the use of system or input based assessment criteria which, to a large extent, flies in the face of the growing acceptance of the importance of animal or output based measures as being genuinely reflective of animal welfare. Fourth, it runs the risk of not always being supported by animal science and of being prone to more affective, anthropomorphic and consumer-driven assumptions of welfare. Finally, there is some concern that a system-based labelling scheme might divert attention away from certain welfare issues that are generic in certain forms of animal husbandry but that are otherwise masked by other 'positive' system attributes (such as, for example, degrees of lameness in free-range sheep flocks or dairy herds). This latter issue could perhaps, though, be addressed either by the judicial use of output measures in support of system based criteria or by the incorporation of selective system or resource based measures into assessment protocols such as that developed by Welfare Quality® (Buller and Roe, 2008).

Over the last 10 or so years, farm animal welfare has grown considerably as a societal issue, backed by a continually expanding legislative and regulatory base, supported by the generalization of assurance standards and, to some extent, driven by an increasingly segmented food market. For many consumers, animal products that do not conform to minimal legislative requirements (or which come from systems that while 'legal' nonetheless fall below societally acceptable standards) should simply not be available for purchase. However, while many are content to let food chain actors and retailers act on their behalf to ensure that such standards are met, a growing number seek a greater commitment to the health and welfare of food animals. It is through a dynamic combination of legislative and regulatory powers, dynamic market segmentation, social movements and innovative animal welfare science that this commitment can be addressed and the lives of Europe's farm animals move a step closer to being 'worth living'.

References

Balsevich, F., Berdegué, J., Flores, L., Mainville, D. and Reardon, T. (2003). Supermarkets and produce quality and safety standards in Latin America. American Journal of Agricultural Economics, 85(5), 1147-1154.

Brambell Committee (1965). Report of the Technical Committee to enquire into the welfare of animals kept under intensive livestock husbandry systems. Her Majesty's Stationery Office, London, UK.

Brooklyndhurst (2010). Are labels the answer? Report to Defra, Brooklyndhurst, London, UK.

Buller H. (2009). Strategies for the implementation of the Welfare Quality project output. In: Kjærnes, U., Bock, B.B., Higgin, M. and Roex, J. (eds.) Farm animal welfare within the supply chain: regulation, agriculture, and geography, Welfare Quality® Report no. 8, Cardiff University, Cardiff, UK, pp. 245-262.

Buller, H and Cesar, C. (2006). Retailing animal welfare food products in France. Report 1.2.3. Welfare Quality, University of Exeter, UK and University of Wageningen, Wageningen, the Netherlands.

Buller, H. (2010). The marketing and communication of animal welfare: a review of existing tools, strategies and practice. European Animal Welfare Platform, Brussels, Belgium.

Buller, H. and Cesar, C. (2007). Eating well, eating fare: farm animal welfare in France. International Journal of Sociology of Food and Agriculture, 15(3), 45-58.

Buller, H. and Roe, E. (2008). Certifying welfare: integrating welfare assessments into assurance procedures: a European perspective: 25 key points. Welfare Quality Report 17, University of Cardiff, Cardiff, UK.

Buller, H. and Roe, E. (2010). Welfare Quality Report, 15, Certifying quality: negotiating and integrating welfare into food assurance. University of Cardiff/Welfare Quality, Cardiff, UK.

Buller, H. and Roe, E. (2012). Commodifying welfare. Animal Welfare, 21, 131-135.

Burch, D. and Lawrence, G. (eds.) (2007). Supermarkets and agri-food supply chains: transformations in the production and consumption of foods. Edward Elgar, Cheltenham, UK.

Busch L,. (2010). Can fairy tales come true ? The surprising story of neoliberalism and world agriculture. Sociologia Ruralis, 50(4), 331-351.

Carson, R. (1964.) Foreword. In: Harrison, R. (ed.) Animal machines. Vincent Stuart, London, UK, pp. vii-xxii.

CIWF (2002). Farm assurance schemes and animal welfare: Can we trust them? CIWF, London, UK.

Commission of the European Communities (2009). Options for animal welfare labelling and the establishment of a European Network of Reference Centres for the protection and welfare of animals COM(2009) 584 final. Brussels, Belgium, 10 pp.

Eurobarometer (2005). Attitudes of consumers towards the welfare of farmed animals. Report 229, Brussels, Belgium.

Eurobarometer (2007).Attitudes of consumers towards the welfare of farmed animals. Report 270, Brussels, Belgium.

Evans, A. and Miele, M. (2007). Consumers' views about farm animal welfare. Welfare Quality Report Series No.4. Part I: National Reports based on Focus Group Research. Cardiff University/Welfare Quality, Cardiff, UK.

Evans, A. and Miele, M. (2008). Welfare Quality Report Series No.5: Comparative Focus Group Report. Cardiff University /Welfare Quality, Cardiff, UK.

Farm Animal Welfare Council (2001). Interim report on the animal welfare implications of farm assurance schemes. FAWC, London, UK.

Farm Animal Welfare Council (2006). Report on welfare labelling. FAWC, London, UK.

Farm Animal Welfare Council (2011) Education, knowledge application and communication in relation to farm animal welfare, FAWC, London., UK.

Farm Animal Welfare Forum (2010). Labelling food from farm animals., FAWF, Godalming, UK.

Farm Sanctuary (2009). The truth behind the labels: an assessment of product labeling claims, industry quality assurance guidelines and third-party certification standards. Farm Sanctuary, New York, NY, USA.

Fordyce, P. (2011). Welfare and quality assurance schemes. Veterinary Record, 169, 344.

Frewer, L.J., Kole, A., Van de Kroon, S.M.A. and De Lauwere, C. (2006). Consumer attitudes towards the development of animal-friendly husbandry systems. Journal of Agricultural and Environmental Ethics, 18(4), 345-367.

Fulponi, L. (2006). Private voluntary standards in the food system: the perspective of major food retailers in OECD countries, Food Policy, 31, 1-13.

GHK (2010). Evaluation of the EU policy on animal welfare and possible policy options for the future: Report to DG Sanco. GHK, Clerkenwell, UK.

Godley, A. and Williams, B. (2007). The chicken, the factory farm and the supermarket: the emergence of the modern poultry industry in Britain. Economics & Management Discussion Papers em-dp2007-50, Henley Business School, Reading University, Reading, UK.

Grabosky, P.N. (1994). Green markets: environmental regulation by the private sector. Law & Policy 16, 419-448.

Harrison, R. (1964). Animal machines. Vincent Stuart, London, UK.

Henson, S.J. and Reardon, T. (2005). Private agri-food standards: implications for food policy and the agri-food system. Food Policy, 30, 241-253.

Hobbs, J. (1996). A. transaction cost analysis of quality, traceability and animal welfare issues in UK beef retailing. British Food Journal, 98(6), 16-26.

Horgan, R. and Gavinelli, A. (2006). The expanding role of animal welfare within EU legislation and beyond. Livestock Science, 103(3), 303-307.

Hughes, A. and Reimer, S. (2004). Geographies of commodity chains. Routledge, London, UK.

Hughes, A., Wrigley, N. and Buttle, M. (2010). Ethical campaigning and buyer-driven commodity chains: transforming retailers' purchasing practices? In: Goodman, D., Goodman, M. and Redclift, M. (eds.) Consuming space: placing consumption in perspective. Ashgate, Godalming, UK, pp. 123-146.

IGD (2007). Consumer attitudes to animal welfare: a report for freedom food. IGD, Watford, UK.

Kjaernes, U. (2005). Consumer concerns for farm animal welfare: a theoretical framework. In: Roex, J. and Miele, M. (eds.) Welfare Quality Report 1: Farm animal welfare concerns. University of Cardiff/Welfare Quality, Cardiff, UK, pp. 53-80.

Kjærnes, U. and Lavik, R. (2008). Opinions on animal welfare and food consumption in seven European countries. In: Kjærnes, U., Bock, B.B., Higgin, M. and Roex, J. (eds.) Farm animal welfare concerns; consumers, retailers and producers, Welfare Quality® Reports no. 7, Cardiff University, Cardiff, UK.

Kjærnes, U., Bock, B., Higgin, M. and J. Roex, J. (eds.) (2009). Welfare Quality reports no. 8. Farm animal welfare within the supply chain: regulation, agriculture, and geography. Cardiff University/Welfare Quality, Cardiff, UK.

Kjorstad, I. (2005). Consumer concerns for farm animal welfare: literature reviews. In: Roex, J. and Miele, M. (eds.) Welfare Quality Report 1: Farm animal welfare concerns. University of Cardiff/Welfare Quality, Cardiff, UK, pp. 3-51.

Lockie, S., Lyons, K., Lawrence, G. and Halpin, D. (2006). Going organic: mobilizing networks for environmentally responsible food production. CAB international, Wallingford, UK.

Marsden, T., Flynn, A. and Harrison, M. (1998). Creating competitive space: exploring the political maintenance of retail power. Environment and Planning A, 30, 481-498.

Mayfield, L.E., Bennett, R.M., Tranter, R.B. and Wooldridge, M.J. (2007. Consumption of welfare-friendly foods in Great Britain, Italy, Sweden and how it may be influenced by consumer attitudes to and behavior towards animal welfare attributes. International Journal of Sociology of Food and Agriculture, 15(3), 59-73.

Miele, M. (2010). Report concerning consumer perceptions an attitudes towards farm animal welfare. EAWP, Brussels, Belgium.

Murdoch, J. (2005). Retail structures. In: Roex, J. and Miele, M. (eds.) Welfare Quality Report 1: Farm animal welfare concerns. University of Cardiff/Welfare Quality, Cardiff, UK, pp. 83-114.

Prickett, R.W., Norwood, F.B and Lusk, J.L. (2010). Consumer preferences for farm animal welfare: results from a telephone survey of U.S. households. Animal Welfare, 19, 335-347.

Roe, E. and Higgin, M. (2008). European meat and dairy retail distribution and supply networks: a comparative study of the current potential markets for welfare friendly foodstuffs in six European countries. In: Kjærnes, U., Bock, B.B., Higgin, M. and Roex, J. (eds.) Farm animal welfare concerns; consumers, retailers and producers, Welfare Quality® Reports no. 7, Cardiff University, Cardiff, UK, pp. 129-256.

Roe, E. and Marsden, T. (2007). Analysis of the retail survey of products that carry welfare claims and non-retailer led assurance schemes whose logos accompany welfare claims. In: Kjaernes, U., Miele, M. and Roex, J. (eds.) Attitudes of consumers, retailers and producers to farm animal welfare, Welfare Quality® Reports no. 2, Cardiff University, Cardiff, UK, pp. 33-70.

Roe, E., Buller, H. and Bull, J. (2011). The performance of farm animal assessment. Animal Welfare, 20, 69-78.

Roex, J., and Miele, M. (2005). Welfare Quality reports no.1. Farm animal welfare concerns: consumers, retailers and producers. Cardiff University/Welfare Quality, Cardiff, UK.

RSPCA (2006). Everyone's a winner: how rearing chickens to higher welfare standards can benefit the chicken, the farmer, the retailer and the consumer, RSPCA. Horsham, UK.

University of Hertfordshire (2010). Effective approaches to environmental labelling of food products. Final Research report for DEFRA, Project FO0419. University of Hertfordshire, Hertfordshire, UK.

Vanhonacker, F. and Verbeke, W. (2009). Buying higher welfare poultry products? Profiling Flemish consumers who do and do not. Poultry Science, 88, 2702-2711.

Vanhonacker, F., Verbeke, W., Poucke, E. and Tuytten, F.A.M. (2007). Segmentation based on consumers' perceived importance and attitude toward farm animal welfare. International Journal of Sociology of Agriculture and Food, 15(3), 91-107.

Vermier, I and Verbeke, V. (2004). Sustainable food consumption: exploring the consumer attitude-behaviour gap. Working Paper 2004/268. University of Ghent, Faculteit Economie and Bedrijfskunde, Ghent, Belgium.

Chapter 4. The Welfare Quality® vision

Harry Blokhuis, Isabelle Veissier, Bryan Jones and Mara Miele

4.1 Introduction

Scientific developments include the appearance of new disciplines as well as an increasing complexity and specialisation in the various existing disciplines and their interrelations. Not only do scientific fields now comprise more and more specialisations but there are also increasing cross-disciplinary links. While this can be considered a logical and valuable progression, as science evolves these sorts of developments may carry a risk of fragmentation and the loss of sufficient critical mass in specialised areas within institutions. Therefore, collaborative efforts and the establishment of international networks are required (Blokhuis, 2009a). Furthermore, the internationalisation (or even globalisation) of agri-business (e.g. cross border sourcing of animals and their products) as well as the complexity of related problems and issues (including animal welfare) undoubtedly require such multi-disciplinary and international collaboration.

At the Lisbon European Council in March 2000, the European Research Area (ERA) was endorsed as a central component of the process of developing a knowledge-based economy and society in the European Union (EU). It was recognised that the issues at stake and the challenges associated with the technologies of the future require European research efforts and capacities that are integrated to a far greater extent than they were at that time. As such the ERA has become the reference framework for research policy issues in Europe. The EU promotes the ERA objectives and strengthens the scientific and technological basis of the Community through for instance the Framework Programmes for research, technological development and demonstration activities. Thus, the EU Commission is the single most influential force on the European research scene today, accounting for 16% of the sum of Member States' civil R&D budgets (European Commission, 2011).

Welfare Quality® was a research project financed under the EU 6th Framework Programme for Research and Technological Development (FP6) and it clearly addressed the ERA concept of integration to achieve important societal and policy objectives. Due to the multidimensionality of the welfare concept welfare science is by definition multi-disciplinary and a variety of methodologies may be applied within disciplines (e.g. Broom, 1996; Fraser and Matthews, 1997; Hughes and Curtis, 1997; Mellor *et al.*, 2009). Furthermore, many different stakeholders (including the general public, consumers, industry, NGOs, etc.) are involved, and they all have their own opinions and interests regarding the animal welfare issue. For these reasons, leading research groups with the most appropriate specialist expertise, both in animal and social sciences, were integrated in one project.

Welfare Quality® was the largest ever European research project on animal welfare. The project began in 2004 and comprised a partnership of 39 institutions in Europe and, since 2006, four in Latin America. The partners were based in 13 European countries as well as Uruguay, Brazil, Chile and Mexico.

4.2 Accommodating the different animal welfare 'drivers'

During the lifetime of Welfare Quality® the original aims of the project (Blokhuis *et al.*, 2003) obviously evolved as new results or developments became apparent and the direction and content of the research were modified accordingly. However, the main drivers underlying the vision and the general aims remained the same during the project's lifetime which was almost six years.

A number of diverse groups, factors, circumstances and developments were influential in driving and guiding the Welfare Quality® project (Blokhuis, 2008; 2009b; Miele *et al.*, 2011). Of these, three crucial external factors were: (1) public concerns and demands; (2) product supply chains and markets; and (3) policy making and regulatory bodies (Blokhuis *et al.*, 2010).

4.2.1 Public concerns and demands

As described in Chapter 2, several major changes in animal farming have taken place from the second half of the 20th century onwards. Production intensified enormously, farms became highly specialised and there were huge increases in the number of animals kept on each farm and in actual production. Housing conditions and management practices also changed profoundly with the appearance of increased mechanisation and other technological developments. Animal production increased in scale and took a much more industry like approach, with quantity often prioritised over quality.

For a variety of different reasons (cultural, attitudinal and commercial) constructive communication between farmers and the people who ultimately eat what is produced was hampered (Kjaernes *et al.*, 2009b). The activities of consumer groups and animal protectionists and, more recently, the effects of crises such as swine fever, BSE, foot-and-mouth disease and avian influenza led to an increased awareness that animal production is more than just an industry and animal welfare assumed much greater importance for the public (both in their roles as consumers and as engaged citizens).

Indeed farm animal welfare has become an increasingly important issue for the European citizen and consumer and they express a clear demand for higher farm animal welfare standards (Eurobarometer, 2005, 2007; Kjaernes and Lavik, 2008). This increasing interest in farm animal welfare is also reflected in a widespread demand for transparent information about animal welfare issues across Europe. However, differences in primary production, processing and distribution as well as in governance structures and public discourse obviously affect this demand and variations between countries are apparent (Kjaernes *et al.*, 2009a). Moreover, the requests for information on farm animal welfare often seem to reflect just a general interest in the issues and the apparent concern is often not clearly expressed through purchase choice for animal friendly products (Kjaernes and Lavik, 2008; Miele and Evans, 2010).

4.2.2 Product supply chains and markets

Supply chains of animal products now focus more and more on delivering good animal welfare as an important attribute of total food quality. In Europe farmers generally consider animal welfare to be an important aspect of farming (Bock, 2009) and they are very motivated to provide good care for their animals within the limits set by the need to ensure the economic profitability of their enterprises. Farmers are also very much aware that they are operating in a competitive market where they have to comply with food retailers' standards if they want to have access to the market. Indeed, animal welfare credentials are increasingly required by retailers in order to strengthen their brand image (Kjaernes *et al.*, 2009b). There is also an increasing recognition that production and specific quality aspects are negatively affected by conditions that harm animal welfare and this clearly jeopardises profitability (Jones, 1998).

Research carried out in Welfare Quality® showed that farmers are in favour of an objective standardised system of assessing farm animal welfare that could be used throughout Europe and preferably worldwide (Bock, 2009). However the farming community also worries about the costs incurred by welfare assessments, welfare improvements and the need to comply with more stringent regulations. The farmers are also anxious that in the end they will be expected to bear such costs (Bock *et al.*, 2010).

On the other hand, more and more producers, retailers and other food chain actors recognise that consumer concerns for good animal welfare represent a business opportunity that could be profitably incorporated in their commercial strategies (Roe and Buller, 2008). Thus, animal welfare is increasingly used, particularly by retailers, as a component of product and supply chain differentiation (Eurogroup for Animals, 2007; Miele *et al.*, 2005). Such differentiation (and creation of markets) may be based on an 'overall' high welfare level; be related to specific welfare aspects; or be 'bundled' or not with other product characteristics, e.g. 'environmental impact', 'global warming' or 'sustainability' (see below).

In general, animal welfare is increasingly used as an important attribute of an overall concept of 'food quality' (Blokhuis *et al.*, 1998; Buller and Cesar, 2007).

4.2.3 Policy making and regulatory bodies

Legislation on animal welfare has a longstanding tradition in many Member States of the European Union (Bennett and Appleby, 2010; Blokhuis *et al.*, 2008). The Protocol on Protection and Welfare of Animals annexed to the European Community Treaty in 1999 (the Treaty of Amsterdam amending the Treaty on the European Union) was a milestone for the development of the Community's animal welfare policy. This Protocol spells out the obligation to pay full regard to the welfare of animals as sentient beings when formulating and implementing Community policies. The Lisbon Treaty reconfirmed in 2007 the legal recognition of animals as 'sentient beings'.

A range of EU Directives and Regulations now specify requirements, conditions and practices with the aim to ensure good animal welfare for different animal species and categories. These cover for instance several aspects of animal housing and husbandry, transport and slaughter. In general though, current EU legislation largely relies on resource based parameters, i.e. specifying the provision of particular resources and practices (prescriptions). This approach is important to guide decisions on the banning of conditions/practices that are widely considered to result in poor welfare, such as certain housing systems (e.g. battery cages for laying hens) or painful procedures. However, reliance on a prescriptive 'resource based approach' has a major drawback in that it does not assess what is finally most relevant, i.e. the welfare status of the animal.

Moreover, an on-going assurance of good animal welfare using prescriptive legislation requires deeper and continuous detailing of housing design and requirements, management procedures, etc., and this could result in very complicated and inflexible legislation. Also, it is often difficult (if not impossible) to define detailed resource measures in such a way that they provide the same protection of animal welfare under the very different farming, societal and climatic conditions that prevail in the various Member States. If resource based rules are too detailed and restrictive they may actually prevent farmers from choosing husbandry systems and practices to their liking or that fit their specific circumstances most, even if these could result in good welfare in that situation. This clearly does not stimulate innovation. Finally, ensuring compliance with such detailed legislation of husbandry conditions and practices would be virtually unachievable.

In the European Commission's Action Plan for Animal Welfare 2006-2010 (European Commission, 2006) it was stated that efforts will be made to incorporate specific measurable animal welfare indicators where available into existing and future EU legislation. A few of these measures (e.g. foot pad lesions and mortality) were included in the 2007 Directive (2007/43/EC) for meat chickens (implemented in 2010).

More recently the European Union Strategy for the Protection and Welfare of Animals 2012-2015 stated that 'subject to an impact assessment, the Commission will consider the need for a revised EU legislative framework based on a holistic approach. In particular, the Commission will consider the feasibility and the appropriateness of introducing science-based indicators based on animal welfare outcomes as opposed to welfare inputs as has been used so far; the Commission will assess whether such a new approach would lead to a simplified legal framework and contribute to improve the competitiveness of EU agriculture'.

4.3 The Welfare Quality® approach

The Welfare Quality® approach was designed to accommodate the above drivers and developments and to respond to their diverse and sometimes divergent requirements. Transparency of the product quality chain and the provision of guarantees in relation to animal welfare were considered major and overarching requirements. These require the development of trustworthy ways of quantifying how production processes affect animal welfare (Blokhuis, 2009b; Blokhuis et al., 1998) and of providing visibility of such processes and effects to all interested stakeholders (citizens/consumers, industry, government, etc.). Therefore, Welfare Quality® aimed to deliver reliable, science-based, on-farm welfare assessment systems that address stakeholder concerns for poultry, pigs and cattle as well as a standardised system to convey welfare measures into clear and understandable product information.

It was also recognised that a concerted European effort in the area of animal welfare should include research designed to identify practical ways of solving some of the main welfare problems in current animal production. Indeed, the assessment tool is designed to drive farmers' attention to the problems that may exist on their farm and it should be coupled with adequate advice in order to help improve the situation. Welfare Quality® scientists agreed that the development of welfare improvement strategies was of great importance and therefore initiated studies in important areas like handling stress, injurious behaviours, lameness, temperament, etc.

Through its integrated European approach, Welfare Quality® was instrumental in providing a firm basis for the European harmonisation of assessment and information systems. Such harmonisation is essential in order to create a level playing field for

European producers and to provide transparent consumer information and marketing. Also, as a possible basis for future legislation, welfare measures need EU wide support and harmonisation.

Welfare Quality® provided instruments (assessment methods and improvement strategies) which can also drive future developments outside the EU. European agriculture embraces diverse physical environments (e.g. cold Nordic countries to warm Mediterranean ones) and different socio-cultural conditions (Kjærnes *et al.*, 2007). The fact that we took this diversity into account means that the instruments developed in Welfare Quality® are likely to be robust and applicable to many other contexts and countries.

Such a harmonised assessment procedure can also be an invaluable tool for testing and evaluating new housing and husbandry systems as well as new genotypes before they are allowed onto the market. By identifying potential risks, such testing would play a critical preventative role.

Thus, the main aims of the Welfare Quality® project were described as:
- to develop a standardised system for the assessment of animal welfare;
- to develop a standardised way to convey measures into animal welfare information;
- to develop practical strategies/measures to improve animal welfare;
- to integrate and interrelate the most appropriate specialist expertise in the multidisciplinary field of animal welfare in Europe.

As we state elsewhere (e.g. Chapters 5 and 6) animal welfare is a multidimensional concept comprising physical and mental health and embracing aspects as diverse as physical comfort, absence of hunger and disease, possibilities to perform motivated behaviour, etc. In Welfare Quality® a primary requirement was that all these different aspects of welfare had to be covered and be stated clearly. These aspects had to reflect what is meaningful to animals, as understood by animal welfare science, and also be agreed upon by the public and other stakeholders in order to ensure that wider ethical and social issues were taken into account. Therefore, in Welfare Quality® we devised ways of establishing dialogue between the project's scientists and the various social constituencies (ordinary citizens/consumers, farmers, breeders, retailers, certification bodies, NGOs, etc.).

4.4 Holistic approach

Varied definitions of animal welfare have been proposed over the last few decades. Some authors regard animal welfare primarily or even solely in terms of the animal's functioning in its environment: bodily health and good production performances are considered indices of good welfare (Hewson, 2003). In this context of satisfactory functioning, 'welfare will be reduced by disease, injury and malnutrition, good welfare will be indicated by high levels of growth and reproduction, normal functioning of physiological and behavioural processes, and ultimately by high rates of longevity and biological fitness' (Broom and Johnson, 1993; Duncan and Fraser, 1997). Other authors define welfare mainly in terms of animal feelings and proposed that animals should not suffer and that they should have positive experiences (Dawkins, 1980; Duncan, 1993). In other words, if an animal feels well then its welfare is alright. A third view focuses on a natural-living argument, i.e. welfare is safeguarded if animals are able to live according to their nature and to perform their full range of behaviours (Rollin, 1981) even if this runs the risk of exposure to inclement weather, predation, food shortage, etc. Clearly, all of these definitions reflect the recognition that welfare is not a unitary concept corresponding to a single physiological or psychological mechanism but rather that it includes many features. In other words welfare is a multidimensional concept (Fraser, 1993, 1995) that embraces:

- freedom from suffering (where suffering refers to negative feelings such as prolonged pain, fear, hunger, thirst...);
- high level of biological functioning (absence of disease, injuries, malnutrition...);
- existence of positive experiences (comfort, contentment, expression of the species-specific behavioural repertoire).

The multidimensional nature of animal welfare requires that all its dimensions should be incorporated in an assessment system. In addition, an animal unit can only be considered welfare-friendly if all welfare aspects are fulfilled. This view emerged strongly from the consultation exercises with the public and it has consistently driven the development of the Welfare Quality® assessment system. Typically animal welfare is viewed as a holistic concept: it emerges from various components but is more than a mere sum of these components (this is further elaborated in Chapter 7).

Welfare Quality® partners developed a way of assessing welfare that aimed to cover all its different aspects. They defined 12 welfare criteria, falling within 4 principles (good feeding, good health, good housing, appropriate behaviour), that are supposed to collectively represent an exhaustive list of requirements for good welfare. The criteria and principles are described in greater detail in Chapter 5.

Specific animal-based and resource-based measures were identified in each animal group in order to verify compliance with the welfare criteria (see Chapter 6). For a given farm or slaughter plant, scores were then calculated to express the extent to which that animal unit complied with each welfare principle. Finally, a synthesis of these principle-scores resulted in the production of an overall evaluation score and a welfare categorisation for a particular farm or slaughter plant. During this process, according to the holistic nature of animal welfare, compensation between criteria or principles was limited. Close collaboration between animal scientists, social scientists and mathematicians and the use of mathematical models and methods derived from decision theory enabled Welfare Quality® to develop reliable methods for the overall evaluation of animal welfare in order to support subsequent decisions in this area. This process is detailed in Chapter 7.

The definitions of welfare principles and criteria were achieved through a process of extensive discussion with the involved stakeholders (i.e. retailers, producers, certifying bodies, NGOs, policy makers) and members of the lay public. This consultation exercise was designed to create a fruitful science-society dialogue around the welfare of farm animals and about the best systems of assessing and monitoring it on farm and at slaughterhouses. The procedure used to establish this dialogue was articulated in several steps at key moments of the Welfare Quality® project (see below).

4.5 Science-society dialogue

The project began with a consultation amongst animal scientists in order to create a provisional list of welfare criteria that combined different scientific perspectives (e.g. biological functioning, emotions, natural living, etc.) about how to approach farm animal welfare and about which aspects of an animal's life should be monitored when attempting to fully describe its quality of life. The underpinning assumptions and the approach to assessing and monitoring animal welfare as well as the initial list of welfare criteria was discussed in focus groups with members of the public in seven European countries (Evans and Miele, 2007, 2008, 2012; Miele and Evans, 2010). This consultation aimed to ensure that the Welfare Quality® protocol would cover the European public's main areas of concern and that it would address issues like legitimacy, trustworthiness, and relevance.

This exercise was paralleled by a consultation with the Welfare Quality® Advisory Committee (comprised of representatives of various stakeholders and interest groups and an ethicist) as well as interviews with farmers, retailers and certifying bodies in six European countries (Bock and Van Huik, 2007; Bock and van Leeuwen, 2005; Buller and Roe, 2008; Roe and Marsden, 2006). The results of all consultations were appraised in dedicated meetings between animal and social scientists about how to best accommodate the diverse concerns of the public (e.g. naturalness, access to the outside) the producers (increasing regulation, bureaucracy and rising costs of production) and retailers (practicality and applicability of science based welfare standards).

Encouragingly, there was good agreement between scientists and the public concerning the welfare criteria. After the initial list of welfare criteria and measures was refined to address stakeholders' concerns, Welfare Quality® scientists proposed a scoring model for converting raw data on welfare 'measures' into meaningful welfare 'scores' and for aggregating the results of the welfare scores. This process (like all processes of evaluation) was, by its nature, bound to ethical choices, e.g. the choice of thresholds between what is considered unacceptable versus acceptable or good, or the decision to allow (or not to allow) for good results on some welfare aspects to compensate for poor results on other aspects (Veissier *et al.*, 2011). These evaluative (ethical) decisions regarding scoring were undertaken on the basis of extensive consultations between animal scientists, social scientists and members of the project's Advisory Committee (see also Chapter 7). Furthermore, they formed a key topic of discussion during the citizen jury exercises (see below).

After the Welfare Quality® protocols for the assessment of animal welfare had been drafted and trialled on a number of poultry, pig and cattle farms in various countries, we undertook in-depth public consultations in the form of citizen and farmer juries in three European countries (the United Kingdom, Norway and Italy). These were designed to gain societal impressions of the assessment protocols as they stood in terms of welfare aspects included or excluded, of how different aspects of welfare were measured, how the measures were combined to produce overall welfare scores and how the assessment system could be best implemented.

During the lifetime of the project, the research results were also presented at three large 'stakeholder conferences' (Brussels, 2005; Berlin, 2007; Uppsala, 2009). Each one was attended by a broad-based international audience of farmers' associations, breeding companies, certification bodies, retailers, NGOs, scientists, members of the EU parliament and the EU Commission, national policy makers, and media representatives (over 250 attendees at each conference). Feedback received from the attendees during these meetings helped us to refine the assessment protocols. The complex of interactive exercises undertaken between science and society are shown in Figure 4.1.

4.6 Focus on animal-based measures

For many years efforts to safeguard animal welfare generally focussed on providing adequate resources, e.g. food, water, shelter, litter, that were expected to safeguard the welfare status of the animals. Indeed, European legislation on animal welfare essentially set minimal norms for different resources. For instance, hens are required to be provided with perches, nest boxes, dust baths, etc. (Council Directive 1999/74/EC). Whether or not the hens use these resources or whether the animals are in

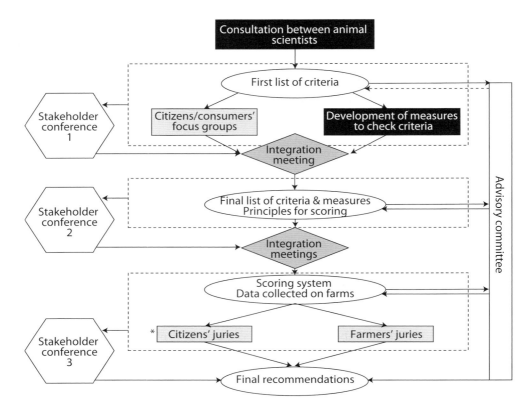

Figure 4.1. Organisation of the dialogue between animal and social scientists, and between science and society during the Welfare Quality® project.

full health fall outside the scope of these directives. Not surprisingly, the sorts of indicators used to check the welfare of animals were often designed to simply assess availability of the resources. Indeed, most of the existing protocols used to monitor animal welfare on farms essentially focus on design- or resource-based measures such as the type of housing, feeding, health plan, etc. Such protocols include the Animal Needs Index (Bartussek, 2001), the Freedom Food schemes (Main *et al.*, 2001), and the decision support tool for sows (Bracke *et al.*, 2002). These assessment protocols generally require a relatively short farm visit of around two hours.

However, although the quality of the animals' living conditions is a key point for their protection, the links between specific design measures and the animals' welfare status are not always clearly understood. Other factors, such as the management and husbandry methods used by the farmer and the genetic background of the animals can profoundly influence the relation between the quality of a resource and the level

Improving farm animal welfare

of welfare achieved. Therefore, the quality of life as perceived by the animal should become the matter of focus (Bracke *et al.*, 1999) and one must determine the animals' mental and physical states. Thus, a protocol designed to assess farm animal welfare, should focus primarily on animal-based indicators, such as the incidence of diseases, whether the animal is in good bodily condition, if it behaves 'normally', if it is unafraid of people, and so on. Using such an animal-based approach Grandin (1994) developed a system to monitor aspects of animal welfare in North American slaughter plants.

Welfare Quality® partners did not wish to make a priori judgements about the welfare offered by different farming systems but rather saw this as a question to be investigated using the measures included in the Welfare Quality® protocols (Welfare Quality®, 2009a,b,c). Therefore, Welfare Quality® chose to focus mainly on the state of the animal, and thus animal-based measures, and not just the nature and quality of its living conditions, although of course these have a large impact on the actual welfare status of the animal.

The 12 welfare criteria mentioned above were defined accordingly, i.e. they focused on how an animal might experience its life; hence criteria such as absence of hunger, absence of thirst, good human-animal relationship, etc. were included (Chapter 5).

For most of the Welfare Quality® criteria it was possible to identify one or several animal-based measures that fulfilled the requirements of validity, repeatability and feasibility. If this was not the case resource-based or management-based measures had to be included in the protocol instead. Resource- or management-based measures are often seen as more feasible because they are normally less time consuming (often needing only one relatively easy recording) whereas animal-based measures require more complex and time consuming observations of several animals.

Collectively, the above considerations resulted in Welfare Quality® protocols that are predominantly animal-based with just a few resource- or management-based measures.

4.7 Welfare improvement strategies and management support

As early as the inception stage the instigators of Welfare Quality® had fully recognised the importance of adopting a broad-brush approach to the pursuit of our predominant vision of safeguarding and improving farm animal welfare. It was clear to Welfare Quality® partners from the start that a concerted effort designed to assess and improve farm animal welfare should include in-depth research aimed at identifying effective and practical ways of eliminating or at least minimising some of the main welfare problems encountered in current animal production. For this reason, the project initiated practical species-specific studies in a number of key areas. The main criteria used to identify the 'target problems' were: (a) the intensity of the problems and numbers of animals affected; (b) whether the problem has a profound effect on any one or more of the Four Principles of Good Welfare identified in the Welfare Quality® project; (c) if alleviation of the problem was likely to improve welfare, productivity, product quality and profitability; (d) if there was an urgent need for high quality scientific research in that area (see Chapter 8).

In this context the Welfare Quality® studies included: the human-animal relationship and handling stress; injurious behaviour, lameness, temperament, neonatal mortality, etc. The main approaches adopted in these studies focused on environmental modification, selective breeding, and improved management. A number of extremely promising strategies were developed, including a sequential feeding regime of diets varying in protein and energy to reduce lameness in broiler chickens, selection tests and proxy indicators for the genetic reduction of aggression and of neonatal mortality in pigs, recommendations for a husbandry practice to alleviate social stress in beef cattle, identification of key features influencing human-animal relationships and the development of a multi-media training programme for stockpersons, etc. These improvement strategies and associated developments are described in greater detail in Chapter 8 and elsewhere (Jones and Manteca, 2009a,b).

A cyclical process of: welfare assessment, in-depth feedback of results and advice, practical improvement strategies, reassessment of welfare, feedback of information and so on is absolutely essential to stimulate the uptake by and maximise the effectiveness of the Welfare Quality® assessment systems for farmers, advisors, retailers and other stakeholders. It should not just be a matter of assessing the animal units and assigning them a welfare score. Feedback of the detailed results of the welfare assessment to the farmer or slaughter plant manager is not only an integral part of the Welfare Quality® vision but it is also essential for the on-going management of their animal unit. When combined with expert advice such information can support the farmer's and abattoir manager's efforts to further improve the welfare, performance and product quality of the animals in their care. Advice on practical and targeted remedial measures is therefore an additional and important incentive to system implementation. The welfare improvement strategies developed in the project represent a critical input to the advisory component of the above cyclical process. Furthermore, by focusing on the 12 criteria and the key welfare problems identified earlier, Welfare Quality® scientists provided the basis for a web-based Technical Information Resource which critically describes the causes and consequences of specific welfare problems as well as their associated practical welfare improvement strategies. Once it has been fully developed the resource will offer different levels of information which can be easily accessed by the reader by clicking on appropriate links.

Although sometimes requiring initial capital and labour investment by the farmers their efforts to improve the animals' quality of life are likely to reduce costs and raise profits as well as allowing the farmers to gain the initiative in the welfare debate (Lawrence and Stott, 2009). Likely benefits include enhanced health, reduced stress and mortality, lower incidence of aggression and other harmful behaviours, improved production, reproduction and product quality, lower veterinary costs, etc. (Blokhuis *et al.*, 2003; Hemsworth, 2003; Jones, 1997). Higher animal welfare standards are also increasingly considered to be a prerequisite to enhancing business efficiency and profitability, satisfying international markets, and meeting consumer expectations (International Finance Corporation, 2006). Indeed, the long-term savings and commercial benefits can far outweigh initial expenditures.

4.8 Practicality and feasibility

Ideally, a practical welfare assessment system should be robust, reliable, trustworthy and affordable; it should also cause as little disturbance to the animals and to the farmer's routines as possible. As it stands though, it could take an assessor between 4 to 8 hours (depending on species) to carry out a complete Welfare Quality® assessment on farm using the current protocols. Not only is this costly but it also limits the number of farms that can be assessed in a given period. Therefore, in order to stimulate uptake by the industry it is essential to reduce the workload and time required while ensuring that the holistic nature of the assessment is retained and that it provides an overall and reliable view of animal welfare.

At first glance this objective seems to demand a very difficult balancing act. However, recent developments as well as some on-going studies may ultimately enable the above requirements to be met. Firstly, for example, software has been developed for the capture of assessment data on farm or at a slaughter plant. This software, which can be used on a computer or on a hand-held 'personal digital assistant' device, will optimise the collection of data in terms of reducing the time needed as well as increasing accuracy. Furthermore, it means that the results of the assessment are generated more rapidly and that their feedback to the farmers for management support is also accelerated. Secondly, recent technological advances could potentially be exploited to automate some of the Welfare Quality® measures. For instance, techniques developed in Precision Livestock Farming such as the use of sensors, sensing devices and real-time modelling (Silva *et al.*, 2009) might potentially complement or even replace some of the time-costly manual measures in the Welfare Quality® protocols. Not only might the application of such automation enhance the effectiveness and time efficiency of the existing welfare assessment protocols but it could simultaneously provide a more continuous registration of data and thereby increase the usefulness of the assessment for management purposes.

The potential identification and validation of proxy indicators, i.e. relatively simple measures that can reliably predict the outcomes of other more complex and/or time consuming ones, could also conceivably improve the assessment protocols and significantly increase the likelihood of their uptake. Another possibility may be the identification and validation of 'sentinel' indicators that are not themselves related to a specific problem but may be used to indicate that there is some problem at the animal unit and that further detailed assessment is required.

The precise nature and frequency of the welfare assessment exercise might also be modified in an effort to increase its adoption. For instance, it may not be necessary to run a full assessment on each farm visit. It might be sufficient for a complete assessment to be followed at a pre-determined interval by a shorter one that simply targets any problem areas that were identified during the first inspection. In this context, it would be beneficial to incorporate risk factor analysis in future studies; this would identify likely problem areas at a particular farm or slaughter plant and help to establish the frequency of visits required for reliable welfare assessment. It may also become possible for the farmer to conduct part(s) of the assessment himself/herself thereby saving time and costs as well as increasing his/her feeling of involvement and investment in the process. These and other potential strategies are discussed in greater detail in Chapter 9.

Similarly, before they can be considered viable, welfare improvement strategies, devices and/or recommendations not only need to satisfy welfare and economic requirements but they must also be practicable. In other words, they have to be safe, affordable, durable and relatively easy to apply by the farmer, slaughter plant manager and/or breeding company. If these requirements are not met the strategy will simply not be implemented (Jones and Manteca, 2009b; Chapter 8).

4.9 A European Union and global reach

The Welfare Quality® project was funded by the European Commission so it was to be expected that the majority of partners were based in Europe. Indeed the partnership initially involved more than 200 scientists who represented 39 institutes and universities in 13 European countries. In 2006 it also gained four partner organisations in Latin America (Mexico, Chile, Uruguay and Brazil). Furthermore, members of the Welfare Quality® advisory bodies and of some intercontinental collaborative research efforts brought input from colleagues in the USA, Canada, Australia and New Zealand. The collective partnership provided a broad sweep of specialist expertise in a number of relevant disciplines ranging from numerous branches of biology to agriculture, mathematics and the social sciences. Stakeholder involvement also offered skills and knowledge from farmer, animal breeding, retail, food service, assurance and certification, veterinary and ethical organisations.

Several reviews of worldwide marketing strategies have shown that producers, retailers and other food chain actors regard farm animal welfare not only as an increasingly important and marketable attribute of an overall concept of food quality but also as a business opportunity in its own right (see also Chapter 3). These 'global' developments were kept firmly in our partners' minds as the project developed. Agriculture in the European Union embraces diverse climates (from warm Mediterranean countries to cold Nordic ones), farming systems and socio-cultural conditions. The fact that we took this diversity into account means that the Welfare Quality® assessment systems and welfare improvement strategies are likely to be robust and applicable to many other contexts and countries. Indeed, some of the welfare assessment protocols have already been successfully tested in Latin America.

References

Bartussek, H. (2001). An historical account of the development of the animal needs index ANI-35L as part of the attempt to promote and regulate farm animal welfare in Austria: an example of the interaction between animal welfare science and society. Acta Agriculturae Scandinavica, Section A, Animal Science Supplementum, 30, 34-41.

Bennett, R. and Appleby, M. (2010). Animal welfare policy in the European Union. In: Oskam, A., Meester, G. and Silvis, H. (eds.) EU policy for agriculture food and rural areas. Wageningen Academic Publishers, Wageningen, the Netherlands, pp. 243-251.

Blokhuis, H.J. (2008). International cooperation in animal welfare: the Welfare Quality© project. Acta Veterinaria Scandinavica, 50(Suppl I), S10.

Blokhuis, H.J. (2009a). Networking and developing international projects. MZEquitation, Skokloster, Sweden. 108 pp.

Blokhuis, H.J. (2009b). The vision of the Welfare Quality project. In: Butterworth, A., Blokhuis, H., Jones, B. and Veissier, I. (eds.) proceedings of 'Delivering animal welfare and quality: transparency in the food production chain'. Welfare Quality© project, Lelystad, the Netherlands, pp. 17-21.

Blokhuis, H.J., Hopster, H., Geverink, N.A., Korte, S.M. and Van Reenen, C.G. (1998). Studies of stress in farm animals. Comparative Haematology International, 8, 94-101.

Blokhuis, H.J., Jones R.B., Geers, R., Miele, M. and I.Veissier (2003). Measuring and monitoring animal welfare: transparency in the food product quality chain. Animal Welfare, 12, 445-455.

Blokhuis, H.J., Keeling, L.J., Gavinelli, A. and Serratosa, J. (2008). Animal welfare's impact on the food chain. Trends in Food Science and Technology, 19, 75-83.

Blokhuis, H.J., Veissier, I., Miele, M. and Jones, R.B. (2010). The Welfare Quality® project and beyond: safeguarding farm animal well-being. Acta Agriculturae Scandinavica A, Animal Science, 60, 129-140.

Bock, B.B. (2009). Farmers' perspectives. In: Butterworth, A., Blokhuis, H., Jones, B. and Veissier, I. (eds.) proceedings of 'Delivering animal welfare and quality: transparency in the food production chain'. Welfare Quality© project, Lelystad, the Netherlands, pp. 73-75.

Bock B.B. and Van Huik, M.M. (2007). Animal welfare, attitudes and behaviour of pig farmers across Europe, British Food Journal, 109(11), 931-944.

Bock, B.B. and Van Leeuwen, F. (2005). Socio-political and market developments of animal welfare schemes. In: Roex, J. and Miele, M. (eds.) Farm animal welfare concerns: consumers, retailers and producers. Welfare Quality® Report no 1. CardiffUniversity, Cardiff, UK, pp. 115-167.

Bock, B.B., Swagemakers, P., Lever, J., Montanari, C. and Ferrari, P. (2010). Farmers' experiences of the farm assessment: interviews with farmers. In: Bock, B. and De Jong, I. (eds.) The assessment of animal welfare on broiler farms. Welfare Quality® Reports no. 18, Cardiff University, Cardiff, UK, pp. 29-70.

Bracke, M.B.M., Spruijt, B.M. and Metz, J.H.M. (1999). Overall animal welfare assessment reviewed. Part 1: Is it possible? Netherlands Journal of Agricultural Science, 47, 273-291.

Bracke, M.B.M., Spruijt, B.M., Metz, J.H.M. and Schouten, W.G.P. (2002). Decision support system for overall welfare assessment in pregnant sows A: model structure and weighting procedure. Journal of Animal Science, 80, 1819-1834.

Broom, D.M. (1996). Animal welfare defined in terms of attempt to cope with the environment. Acta Agriculturae Scandinavica (Section A e Animal Science), 27, 22-29.

Broom, D.M. and Johnson, K.G. (1993). Stress and animal welfare. Chapman and Hall, London, UK.

Buller, H. and Cesar, C. (2007). Eating well, eating fare: farm animal welfare in France. International Journal of Sociology of Food and Agriculture, 15(3), 45-58.

Buller, H. and Roe, E. (2008). Certifying welfare: integrating welfare assessments into assurance procedures: a European perspective: 25 key points. Welfare Quality Report 17, Cardiff University, Cardiff, UK.

Dawkins, M.S. (1980). Animal suffering, the science of animal welfare. Chapman and Hall Ltd., London, UK.

Duncan, I.J.H. (1993). Welfare is to do with what animals feel. Journal of Agricultural and Environmental Ethics, 6, 8-14.

Duncan, I.J.H. and Fraser, D. (1997). Understanding animal welfare. In: Appleby, M.C. and Hughes, B.O (eds.) Animal welfare. CAB International, Wallingford, UK, pp. 19-31.

Eurobarometer (2005). Attitudes of consumers towards the welfare of farmed animals. European Commission, Brussels, Belgium, 138 pp.

Eurobarometer (2007). Attitudes of EU citizens towards animal welfare. European Commision, Brussels, Belgium, 82 pp.

Improving farm animal welfare

Eurogroup for Animals (2007). European retailers and animal welfare. Briefing October 2007. Eurogroup, Brussels, Belgium, 2 pp.

European Commission (2006). Communication from the Commission to the European Parliament and the Council on a Community Action Plan on the Protection and Welfare of Animals 2006-2010. COM 13 final. Brussels, Belgium, 9 pp.

European Commission (2011). Innovation Union Competitiveness Report 2011, Brussels, Belgium, 765 pp.

Evans, A. and Miele, M. (eds) (2007). Consumers' views about farm animal welfare, part I: national reports based on focus group research, Welfare Quality® Reports no. 4, Cardiff University, Cardiff, UK.

Evans, A. and Miele, M. (eds.) (2008). Consumers' views about farm animal welfare. Welfare Quality Reports Series No. 5. Cardiff University, Cardiff, UK, 129 pp.

Evans, A. and Miele, M. (2012) Between food and flesh: how animals are made to matter (and not to matter) within food consumption practices. Environment and Planning D – Society and Space, 30(2), 298-314.

Fraser, D. (1993). Assessing animal well-being: common sense, uncommon science. Food Animal Well-Being, Purdue University Office of Agricultural Research Programs, West Lafayette, IN, USA, pp. 37-54.

Fraser, D. (1995). Science, values and animal welfare: exploring the 'inextricable connection'. Animal Welfare, 4, 103-117.

Fraser, D. and Matthews, L.R. (1997). Preference and motivation testing in animal welfare assessment. In: Appleby, M.C. and Hugher, B.O. (eds.) Animal welfare. CAB International, Wallingford, UK, pp. 159-173.

Grandin, T. (1994). Euthanasia and slaughter of livestock. Journal of the American Veterinary Medical Association, 204, 1354-1360.

Hemsworth, P.H. (2003). Human–animal interactions in livestock production. Applied Animal Behaviour Science, 81, 185-198.

Hewson, C.J. (2003). What is animal welfare? Common definitions and their practical consequences. Canadian Veterinary Journal, 44(6), 496-499.

Hughes, B.O. and Curtis, P.E. (1997). Health and disease. In: Appleby, M.C. and Hughes, B.O. (eds.) Animal welfare. CAB International, Wallingford, UK, pp. 109-125.

International Finance Corporation. (2006). Animal welfare in livestock operations. Good Practice Note no. 6. International Finance Corporation, Washington, DC, USA, 24 pp.

Jones, R.B. (1997). Fear and distress. In: Appleby M.C. and Hughes B.O (eds.) Animal welfare. CAB International, Wallingford, UK, pp. 75-87.

Jones, R. B. (1998). Fear and distress. In: Appleby, M. C. and Hughes, B. O. (eds.) Animal welfare. CAB International, Wallingford, UK, pp. 75-87.

Jones, R.B. and Manteca, X. (2009a). Best of breed. Public Science Review, 18, 562-563.

Jones, B. and Manteca, X. (2009b). Developing practical welfare improvement strategies. In: Keeling, L.M. (ed.) An overview of the development of the Welfare Quality® project assessment systems. Welfare Quality Reports 12, Cardiff University, Cardiff, UK, 57-65.

Kjærnes, U., Bock, B. and Miele, M. (2009a). Improving farm animal welfare across Europe: current initiatives and venues for future strategies. In: Kjærnes, U., Bock, B.B., Higgin, M. and Roex, J. (eds.) Farm animal welfare within the supply chain: regulation, agriculture, and geography, Welfare Quality® Report no. 8, Cardiff University, Cardiff, UK, pp. 1-69.

Kjærnes, U., Bock, B., Higgin, M. and Roex, J. (eds.) (2009b). Farm animal welfare within the supply chain: regulation, agriculture, and geopgraphy. Cardiff University Press, Cardiff, UK, 280 pp.

Kjærnes, U., Miele, M., and Roex, J. (2007). Attitudes of consumers, retailers and producers to animal welfare. Welfare Quality Reports no. 2. Cardiff University, Cardiff, UK, 183 pp.

Kjærnes, U. and Lavik, R. (2008). Opinions on animal welfare and food consumption in seven European countries. In: Kjærnes, U., Bock, B., Roe, E. and Roex, J. (eds.) Consumption, distribution and production of farm animal welfare opinions and practices within the supply chain, Welfare Quality Reports no. 7, Cardiff University, Cardiff, UK, pp. 1-126.

Lawrence, A. and Stott, A.W. (2009) Profiting from animal welfare: an animal-based perspective. Journal of the Royal Agricultural Society of England, 170, 40-47.

Main, D.C., Webster, A.J.F. and Green, L.E. (2001). Animal welfare assessment in farm assurance schemes. Acta Agriculturae Scandinavica, Scandinavica, Animal Science. Sect A. Suppl., 30, 108-113.

Mellor, D.J., Patterson-Kane, E. and Stafford, K.J. (2009). The sciences of animal welfare,. Wiley-Blackwell, Oxford, UK, pp. 3-12.

Miele, M., Murdoch, J. and Roe, E. (2005). Animals and ambivalence, governing farm animal welfare in the European food sector. In: Higgins, V. and Lawrence, G. (eds.) Agricultural governance. Routledge, London, UK, pp. 169-185.

Miele, M. and Evans, A. (2010). When foods become animals, ruminations on ethics and responsibility in care-full spaces of consumption. Ethics, Place and Environment, 13, 58-78.

Miele, M., Veissier, I., Evans, A. and Botreau, R. (2011). Animal welfare: establishing a dialogue between science and society. Animal Welfare, 20, 103-117.

Roe, E. and Marsden, T. (2006) A comparative assessment of the market for welfare-friendly foodstuffs across 6 European Countries. In: Kjarnes, U., Miele, M. and Roex, J. (eds.) (2007) Farm animal welfare concerns: consumers, retailers and producers. Welfare Quality® Reports no. 2, Cardiff University, Cardiff, UK.

Roe, E. and Buller, H. (2008). Marketing farm animal welfare. Welfare Quality factsheet. Wageningen, the Netherlands: Welfare Quality, 2 pp.

Rollin, B.E. (1981). Animal rights & human morality. Prometheus Books, Amherst, NY, USA, 400 pp.

Silva, M., Exadaktylos, V., Ferrari, S., Guarino, M., Aerts, J.-M. and Berckmans, D. (2009). The influence of respiratory disease on the energy envelope dynamics of pig cough. Computers and Electronics in Agriculture, 69(1), 80-85.

Veissier, I., Jensen, K.K., Botreau, R. and Sandøe, P. (2011). Highlighting ethical decisions underlying the scoring of animal welfare in the Welfare Quality® scheme. Animal Welfare, 20, 89-101.

Welfare Quality® (2009a). Welfare Quality® assessment protocol for cattle (fattening cattle, dairy cows, veal calves). Welfare Quality® Consortium, Lelystad, the Netherlands, 182 pp.

Welfare Quality® (2009b). Welfare Quality® assessment protocol for pigs (sows and piglets, growing and finishing pigs). Welfare Quality® Consortium, Lelystad, the Netherlands, 114 pp.

Welfare Quality® (2009c). Welfare Quality® assessment protocol for poultry (broilers, laying hens). Welfare Quality® Consortium, Lelystad, the Netherlands, 114 pp.

Chapter 5. Welfare Quality® principles and criteria

Linda Keeling, Adrian Evans, Björn Forkman and Unni Kjaernes

5.1 Introduction

This chapter outlines the science behind the development of the Welfare Quality® assessment systems. We concentrate on the animal welfare science and social science discussions upon which the welfare principles and criteria developed in the project are based, rather than on the actual assessment measures and their integration which are detailed in Chapters 6 and 7, although examples of measures are given to illustrate some of the issues that arose during the above process (Forkman, 2009; Keeling, 2009).

From its inception the Welfare Quality® project was committed to developing a new way of assessing farm animal welfare that was both scientifically rigorous and reflected broader public concerns. Within the project social scientists worked alongside animal scientists in order to gain a deeper and more broad-based understanding of societal concerns about farm animal welfare and to help foster science-society dialogue around these issues based on fact rather than supposition (Miele *et al.*, 2011). In some ways this was fairly straightforward, for example a societal concern about injuries to farm animals can be easily matched to the definitions of animal welfare prevalent among scientists. However, in other ways this dialogue was quite challenging because it raised fundamental questions not only concerning what constitutes 'farm animal welfare' but also about the nature and relevance of dialogues between experts and members of the public around these kinds of issues. For example, how can we foster a constructive dialogue between groups with very different levels of perception, understanding, experience and expertise? How also can we take into account the opinions of non-experts within the increasingly technical fields of modern farming and farm animal welfare?

In the Welfare Quality® project the term animal welfare primarily came to be a description of the quality of an animal's life as it is experienced by an individual animal (Bracke, 1999). It is regarded as a dynamic state that is reduced when the animal experiences states of pain, fear and suffering and enhanced when animals experience pleasurable states (Broom, 1996; Duncan, 1996). When used in this way the term 'animal welfare' refers to the actual state of an animal rather than to the ethical obligations that people have to care for animals. However, ethical decisions underlie judgements about animal welfare and therefore such decisions also underpin the structure of the Welfare Quality® assessment systems; especially concerning the weighting of different criteria and the establishment of the cut off points/thresholds between the various welfare categories – acceptable, good, enhanced, excellent (Veissier *et al.*, 2011).

Societal concern over animal welfare has increased significantly in Europe since the 1960s and it has been mainly focused on farm animals. Farm animal welfare is now relevant to a wide range of stakeholders and many people have developed their own personal views on what is and what is not important to measure or consider. It is generally agreed that an animal welfare assessment system should be based on sound scientific knowledge of animal welfare, but for the system to be widely accepted it also has to satisfy public, industry and political views of animal welfare as well as the ethical aspects. In other words, the assessment system should address all valid areas of concern and not just those of interest to researchers. As referred to earlier in this book, concerns about animal welfare have been classified into three broad types: (1) concerns about the animal's feelings or emotions; (2) concerns about the animal's ability to function adequately in a biological sense; and (3) concerns about the naturalness of how the animal lives (Fraser *et al.*, 1997). These 'concerns' are usually reflected in descriptive definitions of animal welfare, e.g. the World Organisation of Animal Health (OIE) defines an animal as being in a good state of welfare if (as indicated by scientific evidence) it is healthy, comfortable, well nourished, safe, able to express innate behaviour, and not suffering from unpleasant states such as pain, fear, and distress. Good animal welfare requires disease prevention and veterinary treatment, appropriate shelter, management, nutrition, humane handling and humane slaughter/killing. Animal welfare refers to the state of the animal (OIE, 2008).

One of the most widely used sets of recommendations concerning animal welfare is contained in the Five Freedoms proposed by the Farm Animal Welfare Council (FAWC) in the UK. These freedoms define ideal states to aim towards and they include: (1) freedom from hunger and thirst – by ready access to fresh water and a diet to maintain full health and vigour; (2) freedom from discomfort – by providing an appropriate environment including shelter and a comfortable resting area; (3) freedom from pain, injury or disease – by prevention or rapid diagnosis and treatment; (4) freedom to express normal behaviour – by providing sufficient space, proper facilities and company of the animal's own kind; and (5) freedom from fear and distress – by ensuring conditions and treatment which avoid mental suffering (FAWC, 1979).

From the original start up meetings of the project there was a general consensus that the Welfare Quality® assessment system should include these aspects of the Five Freedoms. However as stated above, the FAWC definitions are descriptive or refer to ideal states. To be useful in practice these definitions have to be translated into more operational ones with measurable criteria. Since it is generally agreed that welfare is a multidimensional concept a list of mutually exclusive domains that could be evaluated was constructed. The process by which Welfare Quality® researchers came to the twelve welfare criteria currently included in the assessment schemes is presented later in this chapter.

Parallel to this philosophical and scientific discussion on what constitutes good welfare and a good quality of life we also considered a number of other questions. These included: (a) should animal welfare be defined primarily in negative terms, such as the avoidance of pain, or in positive terms such as the expression of presumed pleasurable behaviour (like play); (b) should ideas of natural living be central or peripheral to the definition of welfare used; (c) should good feeding imply how the food is provided and not only provision of a healthy diet?

There were also extensive discussions on the type of measures that should be used to determine the extent towards which good welfare was achieved. These can generally be divided into resource-based measures (e.g. space allowance, type of floor, climate control systems, etc.), management-based measures (biosecurity, feeding regimes, handling, etc.) and animal-based measures (e.g. injuries, fear, lameness, etc.). The first two types are generally regarded as 'input' measures and the third one as 'outcome' measures. The decision to use as many animal-based measures as possible for welfare assessment had already been taken based on discussion in the COST Action 846 project (Measuring and Monitoring of Farm animal Welfare) that preceded Welfare Quality® (Blokhuis *et al.*, 2006). The primary argument for focusing on animal-based measures is the fact that they are considered 'close' to the experiences of the animal (see also Chapter 4). A further consideration was the desired ability to compare the welfare of animals kept in different types of farming systems; which again argued strongly for the use of animal-based measures. However, both management and resource-based measures were still regarded as important for identifying causes of poor welfare, risks to welfare or as substitute measures in cases where no valid animal-based measures were available.

At that time this emphasis on animal-based measures was counter to most of the common assessment schemes which relied heavily on resource-based measures (see also Canali and Keeling, 2009). Resource-based measures are parameters that can easily be evaluated in that they require comparatively little training of the assessor and usually have high inter and intra-observer repeatability. They are also typically the sorts of parameters used for legislation. While the resources provided determine the physical situation for the animal, management-based parameters are also very important. Major management decisions regarding the animal's life include: how and when it is fed, moved and mixed with other animals, as well as the use of mutilations, such as routine practices like beak trimming, tail docking or dehorning. Moreover, even apparently minor differences in the way the animal is handled are known to affect welfare. Some management factors can be determined from farm records (if available) or farmer interviews, but the best ways of assessing management would be to directly observe or to film the farmer at his/her work, something that would be impossible to do in an assessment scheme without disturbing and/or altering his/her routine practice. Furthermore, because welfare is a characteristic of the animal, and animals differ in their genetics, early experience and temperament, they may perceive the same environment and management practices in different ways. Collectively, these views reflect the perceived weakness of resource and management-based measures and they underpinned the decision to base the welfare assessment on animal-based measures as much as possible. This decision was confirmed and further strengthened in the early discussions within the Welfare Quality® project.

As indicated above, there are several domains that are generally regarded as areas of concern and of importance for good animal welfare. Based on wide-ranging discussions, the Welfare Quality® scientists agreed that in order to assess the overall welfare status of animals, four main areas of concern should be addressed and four related key questions answered. These questions include: are the animals properly fed and watered, are they properly housed, are they healthy and can they show appropriate behaviour? This development evolved into the four Welfare Quality® principles (Figure 5.1). On closer examination it was clear that each area of concern encompassed several more detailed aspects that needed to be addressed. Thus, two to four welfare criteria were defined within each of the four principles.

An initial list of 10 criteria was later revised to what has since become known as the 12 Welfare Quality® criteria (Figure 5.1). As our knowledge and experience with the Welfare Quality® assessment system is extended and improved in the future, these 12 criteria may further evolve to reflect that increased expertise and information. Rather than limiting our aims to the existing 'knowledge about animal welfare measurements', we included criteria that were regarded as important within the whole welfare framework, even if it was not clear at that time how they might best be measured; a prime example is the criterion defined as 'positive emotional states'.

It was also agreed quite early in the development of the 12 welfare criteria that they should be applicable across all species and all situations. Some criteria may be more critical or more often associated with a particular welfare problem for some species than others. For instance, the range of temperatures for thermal comfort is much narrower in young animals than in adults and likewise, even if good handling at the slaughterhouse is very important, it is still only a very small part of the whole human-animal relationship.

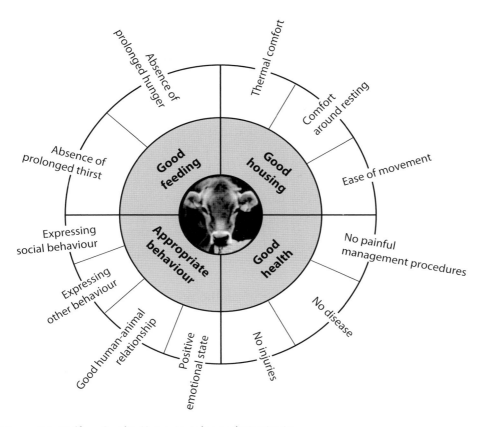

Figure 5.1. Welfare Quality®'s 4 principles and 12 criteria.

One very important requirement that had to be satisfied before any measure could be included in the assessment was that it had to be practical as well as robust and reliable. Aspects related to the feasibility of measures, as well as their reliability and validity are addressed in more detail in Chapter 6.

Measures that could be considered to reflect the welfare of the animals while on farm could also be gathered at other time points. For example, for some measures (e.g. pneumonia in fattening pigs, ascites in broilers), the best opportunity to assess the welfare of the animal when it was at the farm is actually when it has passed through the slaughterhouse, i.e. by post-mortem examination.

A final point before proceeding to the next section, which takes up more specific examples and issues that arose during the development of the assessment system, concerns how the principles and criteria should be worded. Good health and appropriate behaviour are animal-based terms, whereas good feeding and good housing appear to be management or resource-based terms. One could also propose that the principle should actually be called 'Good feeding and provision of water', since thirst is also included. Although terminology in this context was discussed at length the eventual decision reflects a compromise between the precision of language (to depict consideration of the broad area covered by each principle) and the ease of communication. The need for short titles that were easy to communicate to a wide audience was an important consideration during the development of the principles and the assessment system. On the other hand, it was relatively easy to develop short (animal-based) names for the criteria that also reflected the direction in which good welfare could be best achieved by using the terms like 'absence of' 'no' and 'good' together with words that themselves are positively loaded e.g. 'ease of' 'expression of' even though it was recognised that, just like the term 'freedom from' that is used in the Five Freedoms, these ambitions will never be completely realised (Botreau *et al.*, 2007).

5.2 Discussion within Welfare Quality® on the principles and criteria

In this section we discuss some of the issues that arose during the process by which we arrived at the final 12 criteria. This process involved consultation with both the Advisory Committee (stakeholders) and the Scientific Board (scientific experts) of Welfare Quality® and extensive discussions between the animal scientists and the social scientists within the project as well as members of the public (e.g. in focus groups). These efforts are discussed in greater detail later in this chapter. The full lists of welfare measures proposed for each species are presented in Chapter 6. For present purposes this chapter uses selected examples to illuminate the thinking behind the decision processes within the development of the assessment system concept (see Forkman and Keeling (2009a,b,c) for more detail on the examples used for layers and broilers, for sows, piglets and fattening pigs, and for dairy cattle, beef bulls and veal calves).

5.2.1 Good feeding

The seemingly most straightforward criteria are those related to the provision of feed and water, which all animals need in order to survive. It is also clear that to imply a complete absence of hunger and thirst is unrealistic. Indeed to some extent we can regard hunger and thirst as essential motivations for animals to eat and drink. Thus the concept of *prolonged* hunger or thirst was used. With regard to hunger it is also well known that even if an animal is given sufficient food, so that it is no longer metabolically hungry, the food may be in such a form that it does not satisfy the behavioural need to express foraging behaviour. The food is also often consumed too quickly, with many attendant problems (see below). A typical example of 'inappropriate' feeding is when animals are given small amounts of concentrated food when their physiology and behaviour are adapted to consuming large quantities of roughage. In the long term this may lead not only to digestive problems (recorded in the health principle) but also to boredom or frustration and the development of abnormal and harmful behaviours such as stereotypies, feather pecking, tail biting, etc. Thus, two obvious animal-based measures would be how fat or thin the animal is (reflecting the energy content of the food) and the presence or absence of abnormal behaviour (reflecting the appropriateness of the type of feed and the way in which it is given). Body condition scoring has already been well developed in several species, e.g. cattle and pigs, although there are different scoring systems. Being overly fat is a welfare problem, as is being overly thin. However a dilemma arises as to whether animal-based measures of abnormal behaviour should be included in the principle of 'Good feeding' or in the principle dealing with 'Appropriate behaviour'. It was decided that since abnormal behaviour can result from reasons other than just feeding, it should be a component of the 'Appropriate behaviour' principle. From this example it can be seen that it was sometimes difficult to decide *where* exactly to place a particular criterion since it could potentially be located under one of several principles. In these situations it was important to avoid the possibility of double counting a welfare problem. Since welfare assessment includes *all* criteria the exact placing is ultimately less important in the final overall score than it may appear. However, since it obviously influences the principle score, there was always discussion on how to determine the most appropriate placement.

Hunger, even prolonged hunger, is an acknowledged problem in some production systems, such as those for gestating sows and broiler breeders. However, even when it is generally true that there is 'proper feeding' at the group or herd level, it may not be so at the level of individual animals. Individuals in a group differ in their competitive capability and some may be less able than others to compete for resources such as food, water, etc. This situation led to further discussion and questions including: whether all individuals need to be assessed or if it is enough to evaluate a sample, and whether we should be addressing means or medians for the group or if more attention should be directed towards the bottom quartile of individuals with the worst scores and presumably also the worst welfare. While these questions to some extent affected how the measures were constructed their main importance became clear in our considerations regarding how the different welfare measures should be best integrated to provide an overall welfare score. These issues are discussed in Chapters 6 and 7.

The criterion 'Absence of prolonged thirst' led to new discussions since the measures that are available to test dehydration, such as osmolarity (Pritchard *et al.*,

2006), usually involve taking a blood sample. It was decided that the assessment system should not incorporate measures that would require the assessor to have a particular career background. In other words, the assessor should not need to be a veterinarian, as might be the case if a blood sample was required to determine the level of dehydration, or a trained ethologist, as might be the case if the behavioural observations became too complex. It was also decided that the measure should be of such a type that feedback could be given to the animal unit manager directly after the assessment rather than only when additional analyses had been done. For these reasons, animal-based measures of dehydration were excluded. Therefore, in this case it was clear that resource and or management-based measures (access to sufficient clean water) would have to be used until future research has identified a valid, reliable and feasible animal-based measure.

5.2.2 Good housing

Traditionally good housing has been evaluated based on design criteria such as space, floor type, ventilation, presence of enrichment, etc. In keeping with the Welfare Quality® emphasis on animal-based measures, the principle 'Good housing' consists of the criteria 'comfort around resting', 'thermal comfort' and '['. We will use this principle to discuss another dilemma that needed to be addressed before we could decide which measures were to be included in each of these criteria. This relates to climate, i.e. the fact that where a farm is located will influence the choice of measures.

The Welfare Quality® assessment system was designed to address welfare on farm and at slaughter, and it was intended to be usable in all countries irrespective of their variable climates. In the case of thermal comfort, the resources provided and the management tools used will vary according to the prevailing climate in each country. Ensuring thermal comfort in a warm climate probably implies the need to provide access to shade or a cool location, whereas in a cold climate it necessitates access to shelter and warmth. The practice employed may also vary according to whether the animal is living in that environment for most of its life or if it is a purely temporary situation, for example during transport. Furthermore, thermal comfort is also affected by environmental conditions other than just temperature. For example, a temperature that leads to shivering in one animal may cause sweating in another animal exposed to a different humidity or wind speed. It is exactly for these reasons that animal-based measures are the most reliable because of their flexibility and because of the way they reflect housing and management effects. For instance, simple animal-based measures such as shivering or panting can be used to detect likely cold or heat stress respectively in many species.

5.2.3 Good health

The principle of 'Good health' clearly includes physical injuries as well as the presence of disease. But pain is also an area of concern and in some cases that pain can be related to routine practices such as castration in pigs and beak trimming in laying hens. In the majority of countries these practices are legal, but even legal practices can be carried out in more or less painful ways. An early decision was that a Welfare Quality® protocol could not and should not vary according to the country and its legislation, but at the same time that it should assess animal welfare under the assumption that the farm, transport company or slaughterhouse is adhering to the legislative requirements of that country. Thus, the best option for good welfare in this criterion is to refrain from performing the painful management practice. However, if it is performed, then it should be done so in a way that results in the least pain and discomfort.

Tail biting in pigs results in injuries to the tail which most often can be seen as either an open wound (fresh injury) or a shortened tail (a healed injury). However tail docking will also result in a shortened tail. The dilemma was whether tail docking should be scored separately or not. In the end it was decided to score injuries to the hind quarters, including the tail (thereby including evidence of tail biting), and to score whether the tails were docked or not and, if they were, whether anaesthesia was used. A similar approach was used for beak trimming in poultry where the plumage and skin damage are recorded separately from whether or not the beak was trimmed and, if so, the amount of beak removed. Thus, the overall decision was that management practices resulting in mutilation should be placed in a separate criterion within the 'Good health' principle rather than being subsumed into the 'absence of injuries' criterion.

Lameness can be caused by disease or injury, and possibly even by a management procedure such as hoof trimming that was poorly performed or not performed at all, as in the case of overgrown hooves. Whatever the actual cause, lameness will most surely lead to difficulties in movement and thereby it would be a relevant measure for the 'ease of movement' category. However, for some species, such as cattle, it may be possible to diagnose the cause of lameness and so place it in the most appropriate criterion, whereas for others, such as poultry, the actual cause can usually only be identified post-mortem. As discussed previously, a particularly important underlying rule was that the criteria should be mutually exclusive to avoid double counting of measures and a consequently biased overall assessment. Thus, even if in some individual cases the actual cause of lameness may be more correctly attributed to another criterion, it was decided that the most appropriate location for measures of lameness overall, i.e. when all species and all situations were considered, was under the 'absence of injuries' criterion. Moreover, when scores for criteria and measures are integrated, the consequences of any potential misplacement in specific cases are diluted. It is the actual number of lame animals and the severity of lameness that are the essential issues, not which criterion they are located under. In those cases when the assessment system is being used as part of a management tool, then again it is the incidence or prevalence of lameness that is key, and not where this measure is located in the overall assessment scheme.

Our decisions that the Welfare Quality® assessment system should not necessitate the taking of blood samples or that assessors should not have to be trained veterinarians had clear consequences for the principle related to good health. It is for these reasons that only the ways of evaluating symptoms of ill health are described in the assessment protocol; no diagnoses would be given.

5.2.4 Appropriate behaviour

The final principle is 'Appropriate behaviour'. In the earlier section on feeding we mentioned in part why stereotyped behaviour arising from inappropriate diet was better placed under this last principle. But not all types of abnormal behaviour are placed here. For example, although tail biting is also an abnormal behaviour, because of the resulting injury it is placed under the health principle and the 'absence of injuries' criterion in pigs and measured according to the presence of tail damage. For the same reason skin lesions due to cannibalistic pecking in poultry are also placed in the 'absence of injuries' criterion. On the other hand, plumage damage in poultry is used as a measure for feather pecking and, although it can lead to injury, cannibalism and reduced ability to control body temperature, it was considered most appropriate for inclusion under the social behaviour criterion.

Another issue that was discussed at length was the use of the word 'natural' or 'normal' with regard to individual and social behaviour patterns. There is an on-going debate as to what is natural and what is normal for a fast growing, high producing animal; indeed some commonly used definitions use the word 'natural' while others use the word 'normal' thereby reflecting the confusion. When the principles and criteria were discussed in the Welfare Quality® focus groups, the participants placed great importance on the ability of animals to perform natural behaviour, and there was a clear preference for animals to have the opportunity to go outside into a 'natural' environment (see below). In view of the on-going debate it was ultimately decided to use neither 'natural' nor 'normal' in Welfare Quality® but rather to name the relevant criteria 'Expression of social behaviour' and 'Expression of other behaviour'. This allows the incorporation of, for example, the aim to achieve low levels of aggression in a flock, even though aggression can be considered a natural and normal behaviour in some circumstances. On the other hand it was unanimously considered important for laying hens to be able to express the natural behaviours of nesting and dust bathing, not because it is 'normal' or 'natural' but since hens have been shown to be highly motivated to perform these behaviours. As a concession to the importance placed on daylight and outdoor access by society in general, it was decided that the assessment system should record whether or not the animals have the opportunity to go outside, together with some brief information on the extent of shade and vegetation in the outside run.

The 'Positive emotional states' criterion is of particular interest. This criterion was originally termed 'Absence of fear' but the title was changed following discussions with focus groups of consumers who, as is explained later, assume rightly or wrongly that most of the negative aspects of animal husbandry have already been dealt with and that future focus should be on the provision of good welfare through positive emotional states. However, there had hitherto been little investigation of positive emotional states in farm animals. In the absence of such research and of validated methodologies, a qualitative behavioural assessment approach (QBA) was therefore used for this criterion. When carrying out a QBA the assessor scores his or her evaluation of the group of animals based on a number of adjectives. These adjectives are words that the observers use to describe the animals (e.g. content, happy, distressed, etc.). Using a Principal Component Analysis of the scores given on these adjectives, the loading on two factors, one that runs from positive to negative welfare, and one that goes from activity to passivity are obtained (Wemeldsfelder, 2009).

The following section presents in greater detail the way in which the discussion with society was carried out and how the resultant information contributed to the development of the Welfare Quality® principles and criteria described in this chapter.

5.3 Discussion with society

In this section we first outline the various methods that were employed during the Welfare Quality® project to help understand public opinions about farm animal welfare, to assess societal expectations and reactions to the proposed animal welfare monitoring scheme, and to further develop science-society dialogue around farm animal welfare. Second, we provide a very brief overview of societal views on farm animal welfare and we discuss how these views were informed by different information sources (such as the media) and how they were embedded in different everyday practices, such as shopping for food. Third, we examine some key similarities and differences between scientific and societal understandings of farm animal welfare. Finally, we briefly discuss some of the ways in which animal scientists responded to societal concerns. We show that whilst in part the animal scientists responded by considering and examining whether or not these societal views were justified, they also acknowledged that the assessment of farm animal welfare involves a whole series of ethical decisions that would benefit from wider public consultation.

5.3.1 Engaging with public opinion

Over the course of the Welfare Quality® project several social-scientific techniques were used to probe the views and concerns of different stakeholders (including producers, retailers, NGOs, policy makers and members of the broader public). For example, the opinions of producers were primarily assessed via a series of in-depth qualitative interviews with pig, cattle and chicken farmers (Bock and Van Huik, 2008). The views and concerns of retailers and processors, and the viability of introducing a new Welfare Quality® based welfare assessment system, were assessed in two ways: (a) via a detailed market survey of the current availability of 'welfare-friendly labelled products'[13] and assurance schemes across several European study countries, and (b) via a series of key-informant interviews with supply chain actors (Roe and Higgin, 2008).

Several techniques were also used to assess the opinions and concerns of members of the public. First, an academic literature review was undertaken to collate and integrate previous research regarding societal concerns about farm animal welfare (Kjorstad and Kjaernes, 2005). Second, in-depth qualitative focus group research was conducted across seven European countries: Hungary, Italy, France, United Kingdom, the Netherlands, Norway and Sweden (Evans and Miele, 2007, 2008). In each study country 6 or 7 focus groups were undertaken; each of these included different socio-cultural groups such as urban mothers, rural women, empty nesters, seniors, young singles, politically active citizens and vegetarians and one country-specific group, which varied across countries (e.g. hunters, gourmets, ethnic/religious minorities). Each focus group discussion lasted for approximately two hours and a range of topics was addressed, including: food consumption, preparation and shopping habits; knowledge about farm animal welfare (general and product knowledge); interactions with and perceptions of welfare-friendly products (attributes, barriers, ethical dilemmas); issues of responsibility, agency and trust in relation to farm animal welfare. The second half of the focus group discussions consisted of two science-society dialogue exercises. The first of these assessed participants' spontaneous animal welfare concerns/priorities while the second examined participants' reactions to a list of 10 animal welfare concerns that had initially been proposed by animal scientists working in the Welfare Quality® project (see earlier in this chapter). Both these exercises provided valuable feedback to animal scientists working on the project

---

[13] 'Welfare-friendly labelled' products were taken to include both those labelled products which addressed animal welfare directly (e.g. the RSPCA's 'Freedom Food' label in the UK) and those which addressed animal welfare as part of a broader range of issues (e.g. 'organic' or 'quality' labels).

Improving farm animal welfare

and we discuss in detail how they responded to this information later in the chapter. The third and final technique employed to assess the views of members of the public consisted of a representative population survey that was carried out across the same seven European countries as those that featured in the focus group research (Kjaernes and Lavik, 2008). Data were collected through Computer-Assisted Telephone Interviews (CATI), conducted by TNS Global in the period 12-27 September, 2005. The survey was based on probability samples, consisting of 1,500 people in each country. The items covered by the questionnaire include purchasing and eating practices, opinions on consumption and animal welfare, attitudes and beliefs related to animals and the treatment of animals, as well as a range of socio-demographic background questions. The survey followed up several topics from the focus group interviews, offering data that could be generalised to the whole population in each of the seven countries. For example, the survey yielded information on the extent of public concern over farm animal welfare in general, the welfare of various animal species and the degree to which that was reflected during shopping for animal-based food in the various countries. The survey data also allowed analysis of the significance of trust (in retailers, government, etc.) in determining consumer action.

In addition to the research outlined above, we also conducted a series of citizen juries (in the United Kingdom, Italy and Norway, see Miele *et al.*, 2010). Citizen juries are a relatively new methodology that was developed to engage citizens in complex technical and ethical decision making processes. For example, in the United Kingdom they have been used to probe opinion about topics as diverse as nanotechnology, biogenetics, water management and drug policy. They are usually set up to provide an opportunity for a panel of non-experts (the jury) to gain knowledge about a particular topic (often via interactions with experts, see below). The jury is then asked to comment on certain issues or controversies and to provide feedback from an informed but non-expert perspective. This process of public consultation is often carried out over a period of several weeks, thus enabling researchers to gain a much deeper understanding of citizens' concerns, as well as whether and how these concerns change in response to the information provided by the experts.

Each of the citizen juries that were undertaken for the Welfare Quality® project consisted of 12-15 lay members who were selected so as to be broadly representative of a range of different societal views regarding farm animal welfare. For example, in the United Kingdom the jury members consisted of: 2 vegetarians; 2 consumers on a low budget; 1 health-conscious consumer; 1 environmentally-friendly consumer; 1 halal or kosher consumer; 1 rural woman; 1 parent with young children; and 4 mainstream or 'ordinary' consumers. The jurors met each week over a period of 4 to 5 weeks to listen and react to expert opinion on a variety of issues relating to animal farming and farm animal welfare. In the first session, jurors were introduced to a range of ethical issues surrounding different approaches to farm animal welfare and were provided with practical information about the nature of contemporary animal farming in their country. In the second session, jurors were introduced to the science of farm animal welfare firstly in general terms and then in relation to the approaches adopted in the Welfare Quality® project. This background information fed into sessions 3 to 5, in which members of the jury were asked to provide feedback on the measures included within the Welfare Quality® assessment scheme; to comment on the statistical combination and calibration of different measures used to provide an overall assessment of welfare; and to evaluate different potential mechanisms for implementing the monitoring scheme (e.g. should it be used to provide information to farmers; as a labelling scheme for consumers; to inform legislation, etc.).

Given the breadth and scope of the social-scientific research undertaken we are unable to present all the results in this short chapter. Instead, we focus primarily on the results that arose from the qualitative research with members of the general public and we also include some results from the statistically representative telephone survey of public opinion. Furthermore, we aim to provide the reader with a general overview of public opinion across all the study countries (for a more detailed analysis of some of the national and socio-cultural variations in farm animal welfare concerns, see Evans and Miele, 2008; Miele *et al.*, 2010).

5.3.2 A brief overview of societal views about farm animal welfare

In this section we draw on the results of the research outlined above to provide a brief overview of societal opinion on farm animal welfare. We contend that public opinion on farm animal welfare does not emerge from a vacuum but rather is grounded in particular everyday practices (such as shopping for food and eating) and is informed by things people see or hear in the mainstream media and via the internet. We also believe that this analysis provided essential background information, which informed our subsequent attempts to contrast scientific and societal views about farm animal welfare.

Results from the quantitative survey indicate that throughout all the study countries the majority of people questioned considered farm animal welfare issues to be either 'important' or 'very important' (Table 5.1, see also Kjaernes and Lavik, 2008). The results also show that respondents had very different perceptions of the welfare conditions experienced by different farm animals in their countries (Figure 5.2). Only a small minority of respondents believed that the welfare conditions experienced by dairy cattle in their country were 'poor' or 'very poor' (Table 5.1), whereas there were far greater concerns about the welfare of pigs and chickens (e.g. in France 57% of respondents thought that chickens experienced poor or very poor welfare; Table 5.1).

Table 5.1. A brief overview of some of the societal views about farm animal welfare (Kjaernes and Lavik, 2008).

	Importance of animal welfare (% saying animal welfare is important)	Poor welfare, cattle (% saying that cattle have poor welfare)	Poor welfare, chickens (% saying that chickens have poor welfare)	Consider welfare when shopping (% considering animal welfare when shopping)
Norway	84%	3%	47%	12%
Sweden	83%	5%	40%	25%
United Kingdom	73%	12%	56%	23%
The Netherlands	69%	9%	50%	13%
France	65%	15%	57%	23%
Italy	87%	16%	49%	41%
Hungary	83%	16%	28%	24%

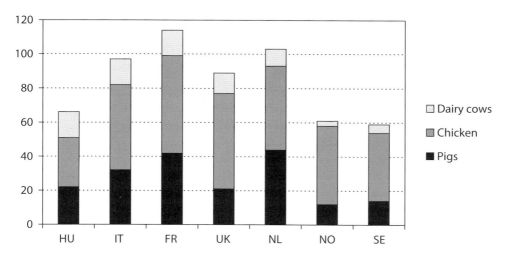

Figure 5.2. Public perception of the welfare of dairy cows, chickens and pigs. The columns indicate the sum of categories 1 ('very poor') and 2 ('poor') per species per country. The specific question asked was: 'In your opinion, how well do you think the welfare conditions are for the following farm animals in [COUNTRY], on a scale of 1-5, where 1 is very poor and 5 is very good?' n=1,500 in each country (HU = Hungary; IT = Italy; FR = France; UK = United Kingdom; NL = the Netherlands; NO = Norway; SE = Sweden) (Kjaernes and Lavik, 2008).

Furthermore, the survey indicates that whilst very high proportions of respondents believed that farm animal welfare was important (in a general, abstract way), far fewer believed that animal welfare was important in the specific context of shopping for food (Table 5.1). Analyses indicate that this distance between general concerns and shopping practices may have a range of explanations, including the view that others (e.g. producers and/or the State) already take responsibility for welfare, the assumption or understanding that problems of farm animal welfare in their own country do not call for consumer action, or that the supply of welfare friendly products is too limited or too expensive.

Results from the focus group discussions confirmed previous suspicions that many contemporary European consumers are detached from the realities of modern farming and that they are poorly informed about *specific* issues of animal biology and contemporary farming practices. However, this is by no means the whole picture, as the research also shows that many participants in the discussions possess detailed understandings of the ethical issues behind farm animal welfare. Furthermore, the majority of focus group participants were able to articulate passionate and well-informed views about what they believed constituted a good quality of life for farm animals and the types of welfare concerns that they believed should be taken into account when monitoring and assessing welfare.

Focus group participants derived their information about farm animal welfare from a number of different sources not only including the media and the internet (see above), but also through direct experiences on farms, through past memories, through analogies based on good human welfare and on their experiences with companion animals. These sources of information and the practices and contexts through which people experienced farm animal welfare issues helped to shape their understandings about what constitutes good farm animal welfare. First, it is fair to say that most members of the public are 'closer to the fork than to the farm' and this in turn shapes their views (Miele and Evans, 2010). For example, many of our focus group participants framed and understood farm animal welfare in terms of the relationship between food safety/quality and farm animal welfare. More specifically, participants believed that factors such as the overuse of medicines and chemicals, stress and the provision of inappropriate ('unnatural') feed have a negative impact on *both* animal welfare and food safety/quality. In short it would seem that for many, but certainly not all of our focus group participants, a concern for farm animal welfare is deeply interwoven with a concern for personal health and wellbeing and for the safety/quality of food products. This observation was also confirmed through the analysis of the survey data. Second, our focus group participants associated a range of already existing certified/assured products with higher animal welfare standards (e.g. organic, free range, outdoor access, and/or quality assured products). Indeed, the labels associated with such products were often cited as an important source of general information about farm animal welfare by participants, especially in Italy, the Netherlands and the United Kingdom and to a lesser extent in the other study countries. Crucially, these labels and the alternative approaches to animal farming that they represented seemed to be exerting a strong influence (alongside some of the other factors mentioned above) over our focus group participants' understandings of what might constitute good farm animal welfare. For example, many focus group participants equated 'organic' with high animal welfare and drew on ideas about organic farming (e.g. relating to feed, outdoor access and naturalism) when discussing animal welfare. In this regard, the participants' reflections were clearly framed by the range of products that were available in shops in different European countries and, in particular, by the fact that broader labelling schemes (such as organic) are far more prominent in many of these countries than specific animal welfare labels[14].

5.3.3 The relative agreement between scientific and societal views

In general, members of the public reacted very positively to the approach to farm animal welfare that was being proposed by animal scientists working on the Welfare Quality® project. They also shared many of the scientists' concerns about farm animal welfare. However, there were certain important differences. Some of these differences reflected substantive distinctions, whereas others were more superficial and reflected differences in terminology or different starting points.

First, our focus group and citizen jury participants were often very critical of high-intensity, 'industrial' farming systems and they believed that these systems could not offer the same level of animal welfare as low intensity systems (e.g. organic or traditional). This view reflected concerns about the amount of space available, the freedom of movement for animals, problems with the types of 'extreme' breeds used in intensive systems and the perceived over-use of medication. Concerns were also raised that farmers would not be able to fulfil their roles as animal carers in 'industrial' contexts. Scientists from the Welfare Quality® project also shared many of these concerns but they approached these issues from an animal-based perspective rather than from an environmental or risk-based perspective. For example, whereas a member of the public might be concerned about the amount of space available per animal, a Welfare Quality® scientist would be likely to view this issue more specifically in terms of the ways in which limiting the amount of space available might thwart various motivated behaviours. However, as a result of this animal-based approach,

---

[14] In making these observations we do not wish to take up a position on the validity of 'organic' inspired versions of farm animal welfare in relation to other alternatives. We are merely making an observation about how the material availability of certain types of products helps to shape public opinion – not only about those products, but also about farm animal welfare more generally.

and in contrast to popular understandings, the Welfare Quality® assessment system does not make *a priori* judgements about the welfare credentials of different farming systems but rather sees this as an empirical question to be investigated. This does not mean that Welfare Quality® scientists think that there are no problems with certain practices and housing system, but rather that they believe that the best ways to understand and deal with the potential problems with such issues is to undertake a detailed animal-based assessment. The results of such an assessment could identify risk factors and guide the development of remedial strategies.

Second, many of the focus group and citizen jury participants highlighted the importance of providing natural environments for farm animals and many of them believed that farm animals should be allowed to live 'as close to a natural life as possible'. Whilst this embrace of naturalism often incorporated overly nostalgic and idealised versions of past farming practices, other elements reflected a more refined understanding of the importance of allowing animals to perform natural/instinctual behaviours, and the benefits of having animals that are 'fit for their environments'. In contrast, Welfare Quality® scientists are far more critical and reserved in their embrace of *environmental* forms of naturalism (i.e. where 'natural' is understood primarily in terms of natural environments). This is because, as discussed above, Welfare Quality® focuses on outcomes (animal-based indicators) rather than just the environment (resource-based measures). Furthermore, Welfare Quality® scientists are generally much more aware of the welfare risks (e.g. predation, disease, bad weather, difficulty in controlling undesirable behaviours (e.g. cannibalism, etc.), that 'natural' environments can pose. However, and in agreement with our research participants, Welfare Quality® scientists did consider the ability to perform several 'natural', or 'species specific' behaviours for which the animals are highly motivated to be an important criterion for achieving good farm animal welfare.

Third, and deeply interconnected with the points above, the majority of focus group and citizen jury participants tended to view animal welfare in terms of resource or input based measures (such as access to a natural environment, amount of space, type of feed, etc.). This was illustrated in the strong preference that jurors expressed for organic schemes. In particular, jurors praised many of the prescriptive principles of organic agriculture, such as specifying that animals must have outdoor access; that the breeds adopted should be suitable for the environment in which the animals will live; that the feed should be organic and non-GMO. Furthermore, whilst jurors acknowledged the potential merits of adopting certain outcome or animal-based measures this approach was not considered to be appropriate for welfare criteria such as 'hunger' and 'thirst'. Indeed, jurors felt that outcome-based measures such as body scores were blunt indicators that were only capable of detecting prolonged periods of hunger rather than shorter episodes and they believed that measures designed to evaluate 'good feeding' should reflect the 'quality' of animal feed in a more direct way (e.g. whether it was natural, if it was genetically modified, etc.). In contrast, Welfare Quality® animal scientists favour outcome or animal-based measures of farm animal welfare (such as health, injuries, behaviour, positive emotion) arguing that these record 'true' welfare status, whereas resource-based measures only identify potential risks to welfare.

Fourth, the concerns of focus group and citizen jury participants tended to focus on positive aspects of farm animal welfare, such as positive emotions, freedom to move and social contact, whereas the criteria proposed by the animal scientists in Welfare Quality® tended to focus on the avoidance of negative aspects of welfare, such as pain, suffering and negative emotions. This was not due to the fact that our jurors and focus group participants placed positive welfare criteria above those dealing with the avoidance of negative situations and states, but rather it was because in their perception issues such as animal suffering should already be dealt with under existing European legislation and that, as such, to be of additional benefit any new European standard for animal welfare should deal with the positive aspects of animals' lives. Moreover, partly as a result of the above, many jurors felt that the Welfare Quality® assessment system would be more suited to measuring welfare in intensive systems (where it would be able to detect instances of poor welfare) but far less suited to detecting and rewarding the very highest levels of welfare (which, for example, many jurors believed existed on organic farms). The difficulty though is the current lack of valid, reliable and feasible measures of positive emotions.

Finally, public concerns were not only limited to the Welfare Quality® welfare criteria in themselves, but also related to how such a welfare assessment scheme might be implemented. The telephone survey as well as the focus group discussions addressed the issue of trust in food chain actors and in the overall system of food production and regulation. A general survey finding was that consumers expressed rather high levels of trust in animal welfare experts. This was also reflected in relatively widespread acceptance of the criteria and considerations developed by the Welfare Quality® scientists, albeit with reservations regarding some elements. Trust in producers and retailers is generally much lower and there is considerable variation, most likely influenced by concrete experiences and specific histories. Behind this, there seems to be an understanding that in the development of standards and assurance schemes, commercial concerns need to be balanced by systems that ensure transparency, third party monitoring, and involvement from animal welfare organisations.

5.3.4 Responding to societal concerns[15]

Differences between scientific and societal views of farm animal welfare were discussed at length within the project. First, in relation to issues of intensive versus extensive systems, most animal welfare scientists did not believe that excellent welfare could only be met in extensive conditions (indeed certain scientists noted that extensive systems can seriously damage welfare in some conditions, for example it may be difficult for range cattle to find water in a drought). Furthermore, scientists pointed out that the term 'extensive' might mean very different things to different people (e.g. outdoor access compared to producing less per unit – where units can be the animal, a piece of land or a worker). One way of reconciling scientific and societal views on this issue would be to better characterise what is defined as 'extensive' versus 'intensive' and to check the results obtained by the various systems by using the Welfare Quality® assessment protocols. If it emerges that some systems that were *a priori* judged positively (e.g. within current prescriptive assurance schemes) do not score well when assessed with the Welfare Quality® protocols, a thorough analysis would be necessary to establish whether the protocols themselves needed to be adjusted or whether the systems that were assumed to be desirable actually turned out to compromise welfare.

Second, in relation to issues of 'natural living', animal scientists recognized that natural environments generally offer the animals more freedom as well as more opportunities to express a wide range of behaviours. Furthermore, experiments using conditioning or maze/runway techniques have shown that animals are motivated to express certain natural behaviours, such as walking (Jensen, 1999; Veissier *et al.*, 2008), social reinstatement (Faure and Jones, 2004; Jones *et al.*, 1999) or social interactions (Patterson-Kane *et al.*, 2002). Hence, access to pasture (for ruminants) or to an outdoor range (for poultry) were added as measures for the criterion 'Expression of other behaviours'. Interactions between animals, exploration of the environment (in pigs), and actual use of an outdoor range were also included in the assessment system.

Third, in relation to whether to focus on input or outcome-based measures the Welfare Quality® scientists decided to prioritise animal-based measures and to only use resource-based measures when direct animal-based measures were not possible or appropriate (e.g. if they were not sensitive enough or they took too long to record). Welfare Quality® scientists also considered jurors' concerns about whether animal-based measures for thirst and hunger were appropriate or sensitive enough. Given the lack of a sensitive animal-based measure of the absence of thirst (see discussion earlier in the chapter), the scientists decided to use resource-based measures for thirst (e.g. checking that animals have access to sufficient water points and to clean water).

---

[15] This section draws on work by Miele *et al.* (2011) and in particular on the inputs of Isabelle Veissier to this paper.

Improving farm animal welfare

However, in relation to hunger the animal scientists pointed out that appropriate feeding is not only a matter of what is provided to animals but that it also depends on the animals themselves and their needs. For example, some animals have higher nutritional requirements, such as lactating dairy cows. Therefore, it was decided that the best way to determine if the animals had been fed appropriately was by observing their bodily condition.

Finally, in relation to the balance between positive and negative measures of welfare, Welfare Quality® scientists agreed that any comprehensive assessment of farm animal welfare should attempt to cover the presence of a positive welfare state as well as the absence of a negative welfare state. Although all criteria include positive and negative dimensions of welfare, these were partially hidden within categories such as 'expressions of behaviour' or 'human-animal relationships'. On a more substantive level, there was a risk that, due to difficulties in identifying appropriate and reliable measures, indicators of positive emotional states might be omitted from the final Welfare Quality® monitoring scheme. However, due in part to the high relevance of positive aspects of animal welfare for European citizens, it was decided to include 'positive emotional state' as one of the 12 welfare criteria and to use Qualitative Behavioural Assessment as a possible way of assessing it.

Thus, one can see how the science-society dialogue took shape within the Welfare Quality® project and how societal opinions were given due consideration when designing a scientifically valid scheme to assess the welfare of farm animals.

5.4 Future implications

It is often difficult to trace back what particular influence determined which specific aspect of the development of the Welfare Quality® principles and criteria. They evolved over the course of the project during which new issues arose and had then to be incorporated in the overall concept. Pragmatic decisions were sometimes made to deal with challenges so that the whole process could proceed to the next step. Regular discussions within the Welfare Quality® partnership (43 partner organisations), targeted meetings where the issues were discussed at depth among partners, as well as substantial debate with and input from the members of the Advisory Committee and Scientific Board (composed of external stakeholders and independent scientific experts, respectively) all contributed to the project's development. Perhaps it is the breadth and depth of input and the gradual evolutionary process that has resulted in the welfare principles and criteria gaining broad and worldwide acceptance. Animal welfare science is a fast growing discipline and it is recognised that the final outcomes of the project, the 4 principles and 12 criteria, the assessment protocols, and the practical improvement strategies, will all need to be updated as new knowledge and technologies emerge. However, the journey, that is to say the process by which the various outcomes were derived, and the knowledge gained during this process has also stimulated new developments in the area of animal welfare assessment.

References

Blokhuis, H.J., Jones, R.B., Veissier, I. and Geers, R. (2006). COST Action 846 'Measuring and monitoring farm animal welfare'. K.U. Leuven R&D, Leuven, Belgium, 48 pp.

Bock, B. and Van Huik, M. (2008). Cattle farmers and animal welfare: a study of beliefs, attitudes and behaviour of cattle producers across Europe. In: Kjaernes, U., Bock, B., Roe, E. and Roex, J. (eds.) Welfare Quality® reports n. 7. Consumption, distribution and production of farm animal welfare. Cardiff University, Cardiff, UK, pp. 257-319.

Botreau, R., Veissier, I., Butterworth, A., Bracke, M.B.M. and Keeling, L.J. (2007). Definition of criteria for overall assessment of animal welfare. Animal Welfare, 16, 225-228.

Bracke, M. (1999). Overall animal welfare assessment reviewed. Part 1: Is it possible? Netherlands Journal of Agricultural Science, 47, 307-322.

Broom, D.M. (1996). Animal welfare defined in terms of attempt to cope with the environment. Acta Agriculturae Scandinavica (Section A – Animal Science), 27, 22-29.

Canali, E. and Keeling, L.J. (2009). Welfare Quality project: from scientific research to on farm assessment of animal welfare. Italian Journal of Animal Science, 8, 900-903.

Evans, A. and Miele, M. (eds.) (2007). Consumers' views about farm animal welfare. Part I: National Reports based on Focus Group Research, Welfare Quality® reports n. 4, Cardiff University, Cardiff, UK.

Evans, A. and Miele, M. (eds.) (2008). Consumers' views about farm animal welfare. Part II: European Comparative Report based on Focus Group Research, Welfare Quality® reports n. 5, Cardiff University, Cardiff, UK.

Duncan, I.J.H., 1996. Animal welfare defined in terms of feelings. Acta Agriculturae Scandinavica (Section A – Animal Science), 27, 29-36.

Faure, J.M. and Jones, R.B. (2004). Genetic influences on resource use, fear and sociality. In: Perry, G.C. (ed.) Welfare of the laying hen, 27th Poultry Science Symposium. CAB International, Wallingford, UK, pp. 99-108.

FAWC, 1979. Farm Animal Welfare Council Press Release. Available at: http://www.fawc.org.uk/pdf/fivefreedoms1979.pdf.

Forkman, B., 2009. Investigating possible measures to include in the assessment systems. In: Keeling, L.J. (ed.) An overview of the development of the Welfare Quality project assessment systems. Welfare Quality® report n. 12. Cardiff University, Cardiff, UK, pp. 9-14.

Forkman, B. and Keeling, L.J. (eds) (2009a). Assessment of animal welfare measures for layers and broilers. Welfare Quality® reports n.9, Cardiff University, Cardiff, UK, 176 pp.

Forkman, B. and Keeling, L.J. (eds) (2009b). Assessment of animal welfare measures for sows, piglets and fattening pigs. Welfare Quality® reports n.10, Cardiff University, Cardiff, UK, 310 pp.

Forkman, B. and Keeling, L.J. (eds) (2009c). Assessment of animal welfare measures for dairy cattle, beef bulls and veal calves. Welfare Quality® reports n.11, Cardiff University, Cardiff, UK, 297 pp.

Fraser, D., Weary, D.M., Pajor, E.A. and Milligan, B.N. (1997). A scientific conception of animal welfare that reflects ethical concerns. Animal Welfare, 6, 187-205.

Jensen, M.B. (1999). Effects of confinement on rebounds of locomotor behaviour of calves and heifers, and the spatial preferences of calves. Applied Animal Behaviour Science, 62, 43-56.

Jones, R.B., Marin, R.H., Garcia, D.A. and Arce, A. (1999). T-maze behaviour in domestic chicks: a search for underlying variables. Animal Behaviour, 58, 211-217.

Keeling, L.J. (2009). Defining a framework for developing assessment systems. In: Keeling L.J. (ed.) An overview of the development of the Welfare Quality project assessment systems. Welfare Quality® report n. 12. Cardiff University, Cardiff, UK, pp. 1-7.

Kjaernes, U. and Lavik, R. (2008). Opinions on animal welfare and food consumption in seven European countries. In: Kjaernes, U., Bock, B., Roe, E. and Roex, J. (eds.) Consumption, distribution and production of farm animal welfare. Welfare Quality® reports n. 7. Cardiff University, Cardiff, UK, pp. 3-117.

Kjorstad, I. and Kjaernes, U., 2005. Consumer concerns for food animal welfare. In: Roex, J. and Miele, M. (eds.) Farm animal welfare concerns: consumers, retailers and producers. Welfare Quality® reports n. 1. Cardiff University, Cardiff, UK, pp. 3-79.

Miele, M. and Evans, A. (2010). When foods become animals, ruminations on ethics and responsibility in care-full spaces of consumption. Ethics, Place and Environment, 13(2), 171-190.

Miele, M., Evans, A. and Higgin, M. (2010). Dialogue between citizens and experts regarding farm animal welfare. Citizen juries in the UK, Norway and Italy. Welfare Quality® reports n. 16. Cardiff University, Cardiff, UK.

Miele, M., Veissier, I., Evans, A. and Botreau, R. (2011). Establishing a dialogue between science and society about animal welfare. Animal Welfare, 20(1), 103-117.

OIE (2008) Resolution on animal welfare. Chapter 7.1. In: Terrestrial animal health code, volume I. OIE, Paris, France. Available at: http://www.oie.int/en/international-standard-setting/terrestrial-code/access-online/.

Patterson-Kane, E.G., Hunt, M. and Harper, D. (2002). Rats demand social contact. Animal Welfare, 11, 327-332.

Pritchard, J.C., Barr, A.R.S. and Whay, H.R. (2006). Validity of a behavioural measure of heat stress and a skin tent test for dehydration in working horses and donkeys. Equine Veterinary Journal, 38, 433-438.

Roe, E. and Higgin, M. (2008). European meat and dairy retail distribution and supply networks: a comparative study of the current and potential markets for welfare-friendly foodstuffs in six European countries. In: Kjaernes, U., Bock, B., Roe, E. and Roex, J. (eds.) Consumption, distribution and production of farm animal welfare. Welfare Quality® reports n. 7. Cardiff University, Cardiff, UK, pp. 129-249.

Veissier, I., Andanson, S., Dubroeucq, H. and Pomiès, D. (2008). The motivation of cows to walk as thwarted by tethering. Journal of Animal Science, 86, 2723-2729.

Veissier, I., Jensen, K.K., Botreau, R. and Sandoe, P. (2011). Highlighting ethical decisions underlying the scoring of animal welfare in the Welfare Quality® scheme. Animal Welfare, 20, 89-101.

Wemelsfelder, F. (2009). The human perception of animal body language: a window into an animal's world? In: Miele, M., Veissier, I., Buller, H., Spoolder, H. And Bock, B. (eds.) Knowing animals, Florence, Italy, p. 79.

Chapter 6. Development of welfare measures and protocols for the collection of data on farms or at slaughter

Isabelle Veissier, Christoph Winckler, Antonio Velarde, Andy Butterworth, Antoni Dalmau and Linda Keeling

6.1 Introduction

One of the main objectives of the Welfare Quality® project was to develop a standardised system for assessing the welfare of animals kept on farms or at slaughter and thus accommodate the main drivers underlying the vision (see Chapter 4). More specifically the need for such an assessment system arose because animal welfare is an important and growing concern for European citizens and of increasing relevance to supply chains and markets as well as policy making and regulatory bodies. It also addresses the fact that European consumers do not feel sufficiently informed about the welfare of farm animals and thereby struggle to take this issue into account when purchasing food or other animal-based products (European Commission, 2007a,b).

Welfare claims are often made regarding animal products across Europe and several schemes have been established to offer welfare-related guarantees to consumers (e.g. the Freedom Food scheme in the United Kingdom and the IKB scheme by the meat industry in the Netherlands (Veissier *et al.*, 2008)). However, these schemes differ in the measures used to assess animal welfare, in the thresholds set to differentiate high versus poor welfare, and/or in the way the information is integrated to form an overall judgement (Botreau *et al.*, 2007). A harmonised assessment system is therefore considered essential to ensure the credibility of welfare claims. Moreover, although welfare improvements do not arise from assessment alone (*you can't fatten a pig by weighing it*), the use of a reliable assessment system is crucial to facilitate improvements to the welfare status of farm animals (Polten, 2007).

A central aim of the Welfare Quality® project was to develop on-farm welfare assessment systems that focus primarily on animal-based measures and that are scientifically sound and feasible (Blokhuis *et al.*, 2003). In this chapter, we describe how we decided which tests and measures were selected to assess farm animal welfare, and we explain the factors underpinning the organisation of visits to the animal units.

6.2 How the welfare measures were developed

6.2.1 Selection of measures

Effective measures and methods are required to assess welfare in all the species and categories covered by Welfare Quality® (i.e. dairy and fattening cattle, veal calves, sows and piglets, finishing pigs, broilers and laying hens). In a first step we selected promising measures from in-depth reviews of the literature (e.g. on-farm and experimental studies) as well as ones based on practical experience (e.g. from existing welfare assessment protocols like the Bristol Welfare Assurance Programme (Main *et al.*, 2007)). Validity was the main criterion used when reviewing potential measures for inclusion in our welfare assessment systems. Validity was defined as the extent to which we can actually measure what we are supposed to, or in other words, the extent to which a measure is meaningful in terms of providing information on the welfare of an animal or a group of animals. The primary focus was on animal-based measures (see also Chapter 4 and 5) which are thought to directly reflect how the animal is faring. Resource- and management-based measures have often been criticised for potentially low validity due to their indirect nature and complex interactions with other resource and management conditions as well as the animal itself which can all result in largely unpredictable relations between such measures and the actual welfare status (Waiblinger *et al.*, 2001). Therefore, resource- and management-based measures were only taken into account when there were no promising animal-based measures available. Two examples are given in Box 6.1. and Box 6.2: one where animal-based measures were found to be more suitable than resource-based ones and one where the opposite case was found.

Reliability of the measures applied (and thus of the results obtained) is crucial for any assessment system. This requirement includes intra-observer reliability which demands that results are largely the same when the same observer repeats assessments using for example. video-clips or pictures; inter-observer reliability, which refers to agreement between the results obtained by two or more observers after they have received reasonable training; and test-retest reliability, i.e. repeated tests with the same subjects are required to yield similar data. A somewhat special but often neglected case of test-retest reliability is the repeatability (consistency) of assessments over time at farm level. This is especially important for measures which are intended to be used for certification purposes in terms of welfare labelling. This means that results must be representative of the longer-term farm situation and not too sensitive to changes in the farming or the weather or in the internal states of the animals as long as the situation has not changed significantly. At the same time, a measure should be sensitive enough to detect variations in welfare state between farms or slaughter plants.

Box 6.1. Example of preference given to animal-based measures.

Criterion 11 of Welfare Quality® focuses on the need for a good human-animal relationship. Such a relationship results in animals that readily approach people rather than avoiding them. This tendency can be measured on farms by measuring the avoidance distance of animals, i.e. how closely can a person approach them before they step away (Lensink et al., 2003; Waiblinger et al., 2006)? This measure can be taken in various situations and at different times of the day. For instance, animals can be approached at feeding; this may help standardise the time of the day when the measure is taken as well as the motivation of the animal to feed; there is also generally sufficient space for the animal to move back from the trough. In pigs, fear of humans can be expressed by panic responses when the observer enters the pen. The inter-observer repeatability of these measures is good and the test is easy to perform on a farm. Importantly, there is evidence that the results are stable over time (Lensink et al., 2003).

The animals' behaviour towards humans depends largely on the attitude and behaviour of humans towards them; potentially influential features include frequency of interactions, quality of interaction (gentle vs. rough), number of people working with the animals, etc. The importance of this interplay has been substantially documented in the different farm animal species (Boivin et al., 2003; Hemsworth and Coleman, 1998; Waiblinger et al., 2006). However, the quality of the human-animal-relationship can depend on many other factors, such as the social environment and the animals' genetic backgrounds (Raussi et al., 2003). This makes it difficult to predict precisely the animals' level of fear of humans or their docility just from information on the farmers' attitude and behaviour. In addition, although it is possible to ask farmers about their attitude towards animals and to observe them during their routine work with animals (e.g. providing them with food, supervising their mating), this appears no less time consuming than observing the animals' avoidance distances. Consequently, it was concluded that it is difficult to determine the true quality of the human-animal relationship from resource- or management-based measures alone. It was thus decided to include the avoidance test and/or the panic test (both animal-based measures) in the Welfare Quality® protocols.

Surprisingly, during the development of Welfare Quality® protocols we found that few welfare indicators had actually been subjected to thorough reliability testing (Engel et al., 2003; Knierim and Winckler, 2009).

Finally, the selected measures should be applicable to different housing systems and at least have the potential to be applied on-farm. These requirements excluded some physiological parameters that need experimental equipment (e.g. heart rate recordings) or laboratory analyses (e.g. hormone assay) as well as complex behavioural tests that could not be integrated into the farm routine (e.g. open-field tests). In terms

Box 6.2. Example of preference given to resource-based measures.

Welfare Quality® Criterion 2 focuses on the absence of thirst. The skin test is often used by vets to detect dehydration especially in young animals. This test consists of measuring the time taken after a pinch for the skin to return to its normal contour. However the validity of this test has been questioned because the time needed for the skin to resume its initial position after being pinched is linked neither to serum osmolarity or packed cell volume (which indicate the level of hydration) nor to drinking behaviour (Pritchard *et al.*, 2006). The appearance of sunken eyes is also used by vets to detect dehydration but it seems that this is only symptomatic of extreme cases of dehydration. Such cases of dehydration do not occur in normal farm practices, except in the presence of pronounced diarrhoea or serious neglect. Thus the measure is not appropriate for the purpose of detecting thirst due to suboptimal provision of water, though it might nevertheless be suitable for detecting dehydration after long transport times.

Plasma osmolarity and haematocrit counts increase when animals are deprived of water (Knowles *et al.*, 1995; Pritchard *et al.*, 2006) but more research is needed to establish if these indicators can help to distinguish a moderate dehydration from a severe one. Additionally, this measure requires collection of a blood sample which is not feasible on farm and could only be easily obtained at slaughter (during bleeding).

In view of the above concerns, it was concluded that neither the skin pinch test nor the observation of sunken eyes are sufficiently valid or sensitive to detect absence of thirst. Hence, recording of resource-based measures such as the number of water points, the flow of water and its cleanliness are currently considered more appropriate for assessing compliance with Criterion 2. If access to water is found to be restricted (because there are not enough drinkers, water flow is insufficient or the water is dirty), it might then be fitting to establish whether this causes the animals to suffer and which of them suffer. This could be done by measuring the osmolarity and haematocrit in some animals that seem in poor condition. Nevertheless the latter measures were not included in the Welfare Quality® assessment system for feasibility reasons.

of the feasibility of the whole assessment protocol, we took the position that it should be possible for a single observer to carry out a farm assessment during a one-day visit.

6.2.2 Testing promising measures

All the measures selected in the first step possessed at least face validity, i.e. they were thought to be valid as judged by experts (Scott *et al.*, 2001). Based on this pre-screening, further evaluation of the measures took place after they had been subdivided into three categories.

The *first category* consisted of measures with less obvious face validity, i.e. those that had not previously been further validated in terms of criterion or construct validity.

Criterion validity is based on the relationship of a measure to another welfare-relevant measure, whereas construct validity is based on the experimental proof that the welfare state is related to the measure in question (e.g. through pharmaceutical intervention). Below we give some examples of measures falling into this category. These include measures of the human-animal relationship in laying hens and measures of exploratory behaviour in pigs and cattle; it was suggested that the latter might conceivably be used as potential indicators of positive state and hence good welfare.

- **Example 1**
  Providing additional human contact to non-caged laying hens by walking through the pen, talking to, touching and individually feeding animals twice daily over a 14-day period significantly reduced their avoidance of people (avoidance distance test) and increased the number of animals that could be touched (touch test) and of those that came close to the observer (stationary person test). Taken together these findings suggested that the tests effectively measure the human-hen relationship (Graml *et al.*, 2008). The results are also consistent with those obtained in caged laying hens (Barnett *et al.*, 1993; Hemsworth *et al.*, 1993). However, the only test that is readily applicable in both cage and non-cage systems was the avoidance distance test (Graml *et al.*, 2009).

- **Example 2**
  The self-rewarding nature of exploration as well as the increase in environmental certainty with the consequent effect of increased predictability and control of the environment is assumed to be associated with positive emotions (Boissy *et al.*, 2007). Thus, measuring exploratory behaviour might be a useful tool in the assessment of animal welfare from the positive emotions point of view. Based on the hypothesis that bulls kept in barren environments (here: pens with fully slatted floors) would explore an unknown object more than those housed under more enriched conditions (here: littered pens), the responses of beef bulls to a simple novel object placed in the feed rack were investigated (Schulze Westerath *et al.*, 2009). However, only slight differences were observed and the test was not sensitive to short-term change in stimulation following a simple enrichment (two lengths of a metal chain and two lengths of rope). Consequently, this test appeared unhelpful for the on-farm assessment of 'positive' welfare state.

- **Example 3**
  The sensitivity of a novel object test designed to measure pigs' (residual) motivation for enrichment was investigated by evaluating their interactions with a novel piece of rope following exposure to different enrichment stimuli (jerrycan, sawdust) (Bracke and Spoolder, 2008). Based on short-term experiments, the results indicated that the novel object test may detect relatively minor differences in

environmental enrichment and may therefore be useful to assess positive welfare through detecting changes in the pigs' motivation to redirect exploratory behaviour. However, the effects of longer-term exposure to environmental enrichment would have to be investigated before such a test can be recommended for inclusion in on-farm assessment protocols.

Exposure to novelty often elicits fear (Jones, 1987; Jones and Boissy, 2011) with behavioural inhibition seen at high levels and cautious investigation at low ones (Jones, 1987, 1997; Salzen, 1979). So, an induced fear state or even a simple lack of interest in the particular novel objects used may conceivably have confounded the tests described in Examples 2 and 3. These considerations illustrate the complexity of emotional interplay and the difficulties likely to be encountered in designing robust but practical tests of welfare. In this situation for example it is therefore essential to check that the stimulus object used during the test can elicit interest without inducing intense and confounding fear reactions.

The *second category* refers to measures that had already been validated (e.g. measures of human-animal relationship in dairy cattle (Waiblinger *et al.*, 2006) or that had high face validity (e.g. lameness in dairy cattle). Lameness is often regarded as the most important welfare problem in dairy cattle due to the pain and discomfort it is likely to cause the affected animals. On-farm lameness assessment is usually based on gait scoring (Flower and Weary, 2006; Winckler and Willen, 2001). More recently, additional construct validation was provided when researchers reported improved gait scores after the application of local anaesthesia (Rushen *et al.*, 2007). Criterion validity was implied by a finding that the frequency of visits by dairy cows to an automatic milking system was related to their locomotory ability. Similarly, the incidence of agonistic interactions in pigs or cattle is generally accepted as a valid indicator of welfare, i.e. frequent and/or intense aggression compromises the animal's welfare. For this type of measure the focus was placed on reliability testing, i.e. inter-observer agreement and consistency over time, and standardisation of the measures rather than on further validation studies. Below, examples of measures that were experimentally validated across species and categories in the project are briefly described. The majority of behavioural measures were subjected to thorough reliability testing across all species. First, testing of inter-observer agreement used either direct observation or the analysis of video clips and pictures. Agreement between observers ranged from acceptable to excellent for several measures of resting behaviours in cattle (Plesch *et al.*, 2010) (see also Table 6.1). Similarly, excellent inter-observer agreement was achieved for a novel object test indicative of general fearfulness in laying hens (Forkman *et al.*, 2009) or the prevalence of oral stereotypies in sows (Courboulay *et al.*, 2009). On the other hand, results for measures of spontaneous behaviour in pigs in the home pen (e.g. agonistic behaviour; Courboulay *et al.*, 2009) or upon arrival at the

Table 6.1. Measures of social behaviour in dairy cattle as an example of how decisions on whether or not to include specific measures in the operational protocols were made based on validity, reliability and feasibility issues.

Measure Incidence of...	Validity[1]	Reliability[2,3]		Feasibility	Inclusion in operational protocol
		Inter-observer agreement	Consistency over time		
Agonistic behaviours					
Head butts (without displacement)	Agonistic behaviours induce unpleasant or stressful experiences and can result in injuries[4,5]	W=0.83-0.92	W=0.74	2 h of net observation time	Yes
Displacements		W=0.85-0.92	W=0.84		Yes
Chasing up (lying area)		Not enough data for calculation of coefficients of reliability		Within a given time frame of 2 h the incidence was too low for reliable recording of separate behaviours	(Yes)
Fighting					(Yes)
Chasing					(Yes)
Total agonistic behaviours		W=0.84-0.97	W=0.76	2 h of net observation time	Yes
Cohesive behaviours					
Social licking	Social licking reduces tension[6,7] and may alleviate poor welfare[8]	W=0.93-0.99	W=0.57	2 h of net observation time	No
Social horning	Horning is a social play behaviour and play is reduced under stress[9]	W=0.59-0.86	W=0.68		No

[1] Consensus validity is used here: experts agreed on the validity of a measure according to their experience. We show here some of the arguments they used to consider that the measure say something about the welfare of the animals.
[2] Laister et al., 2009a,b.
[3] W = Kendall's W.
[4] Albright and Arave, 1997.
[5] Menke et al., 1999.
[6] Sato et al., 1993.
[7] Laister et al., 2011.
[8] Knierim and Winckler, 2009.
[9] Reinhardt and Reinhardt, 1982.

slaughterhouse (Dalmau *et al.*, 2010; Plesch *et al.*, 2010; Scott *et al.*, 2009) indicated a need for increased training of observers and/or revision of the definitions used (this was taken into account in the next phases of the project). Additionally, Principal Component scores derived from Qualitative Behaviour Assessment were shown to be highly repeatable for pigs and poultry (Wemelsfelder and Millard, 2009; Wemelsfelder *et al.*, 2009a) while for cattle, acceptable levels of agreement were only reached after additional training (Wemelsfelder *et al.*, 2009b). Clinical measures were expected to achieve better agreement between different observers but, though less effort was expended in this area, inter-observer reliability of clinical indicators was also checked. Here, for example lameness in sows and finishing pigs, showed high repeatability (Geverink *et al.*, 2009). Consistency of results over time was investigated for all measures of spontaneous behaviour in dairy and beef cattle over 3 repeated farm visits separated by 60 and 120 days (Laister *et al.*, 2009a,b; Plesch *et al.*, 2010) for details see below and Table 6.1. Studies of inter-observer reliability at the slaughterhouse focused on indicators of stunning effectiveness and revealed good to very high agreement between observers for measures such as signs of eye movement or excessive kicking in cattle (Westin *et al.*, 2009), corneal reflex in pigs (Dalmau *et al.*, 2009) and signs of insufficient stunning or wing flapping as indicators of the effectiveness of pre-stunning shock in laying hens (Algers *et al.*, 2009b).

The *third category* comprised all those measures for which validated procedures were already available. For example, body condition score and assessment scores for damage to the integument, such as scratches, wounds, or swellings are well accepted indicators of welfare (although the extent to which different types of lesions/bruising lead to impaired welfare is not yet clear; Knierim and Winckler, 2009), and they have been used quite often in different assessment settings (Regula *et al.*, 2004; Veissier *et al.*, 2004). Data on inter-observer reliability or consistency over time were available for measures like lameness scoring (Engel *et al.*, 2003; March *et al.*, 2007; Winckler and Willen, 2001) and behavioural tests of the human-animal relationship in cattle (Rousing and Waiblinger, 2004; Waiblinger *et al.*, 2007) and considered reliable. For other procedures, e.g. clinical measures of disease, reliability was assumed to be at least satisfactory.

6.2.3 Choice of measures

The selection of measures for inclusion in the operational protocols for on-farm application was based on validity, reliability and feasibility aspects (see above). However, there are no clear-cut scientific criteria for setting the limits for what is an 'acceptable' agreement between or within observers or for test-retest-reliability. Therefore, we adopted previously suggested thresholds (Martin and Bateson, 2007) where correlation coefficients of at least 0.7 were regarded as acceptable for 'important measures that are difficult to assess'. Similarly, the lowest limit of 0.4 for Kappa values (Fleiss *et al.*, 2003) was used. Feasibility issues, such as the time or equipment needed to take the measure, were also taken into account.

In Table 6.1 measures of social behaviour in dairy cattle are taken as an example of how these different aspects were integrated. Regarding agonistic behaviours, face validity was deemed sufficient for their inclusion as promising measures. Extensive behavioural observations in different housing systems and countries revealed that some agonistic behaviours, such as fighting, chasing, or chasing from the lying area, occurred too infrequently to allow reliable recordings within the time frame of a 2-hour net observation time (the latter was set as the upper limit of time that could be spent recording spontaneous behaviour in the course of a one day visit for on-farm assessment). On the other hand, discrete behaviours, such as head butts or displacements, as well as the incidence of total agonistic interactions showed good to very good inter-observer agreement and at least an acceptable consistency over time. We therefore recorded all listed agonistic behaviours during on-farm observations but restricted differentiation of data to head butts, displacements and total agonistic behaviours. Social licking and social horning were initially considered as indicators of cohesive behaviours. However, although social licking has often been described as a potential indicator of positive feelings, we argue that while this association might occur in individuals, we doubt the validity of licking as a herd measure (Knierim and Winckler, 2009). Together with its low consistency over time, this led to the exclusion of social licking as a measure of socio-positive behaviour from the operational assessment protocol. Horning was also excluded because both consistency over time and inter-observer agreement were low; it is also difficult to distinguish social horning ('mock-fighting') from true fights. Table 6.2 shows the final lists of measures included in the 'operational' protocols for all species and categories which were subsequently applied on farms.

Studies on some promising measures of animal welfare at the slaughterplant, such as measures of food deprivation in pigs (Dalmau *et al.*, 2009), distress vocalisations in poultry or high pitch vocalizations in pigs (Algers *et al.*, 2009a), failed to identify valid or feasible indicators of animal welfare (Algers *et al.*, 2009b). Therefore, only those measures with high face validity were selected for the operational protocols. Some of these measures only achieved moderate to low inter-observer agreement in reliability studies (e.g. indicators of stunning effectiveness in cattle (Westin *et al.*, 2009) or behavioural measures of fear during unloading (Dalmau *et al.*, 2010), so special attention was paid to training efforts before applying the protocols. Table 6.3 shows the final lists of measures included in the 'operational' protocols for all species and categories which were subsequently applied at slaughter plants.

Table 6.2. Measures in the operational protocols for the assessment of farms for different species and animal categories which underwent practical application and were then further refined (measures collected at the slaughterhouse but used for the assessment of a farm are in italics).

Good feeding

Absence of prolonged hunger

Dairy cows	Body condition (leanness)
Fattening cattle	Body condition (leanness)
Veal calves	Body condition (leanness); feed supply
Sows/piglets	Body condition (sows leanness); feeding management (sows); age at weaning (piglets)
Finishing pigs	Body condition; feeding management
Laying hens	Feeder space; use, placement and maintenance of resources; modification of resources with intention to improve animal welfare; feeder alarms
Broilers	Feeder space; placement of resources; feeder alarms; *emaciated birds*

Absence of prolonged thirst

Dairy cows	Number and type of water bowls, flow rate, cleanliness, functioning of bowls
Fattening cattle	Number and type of water bowls, functioning of bowls
Veal calves	Number and type of drinkers, functioning of drinkers, timing of water supply, cleanliness
Sows/piglets	Number of drinkers; flow rate
Finishing pigs	Number of drinkers; flow rate
Laying hens	Drinker space; drinker alarms
Broilers	Drinker space; drinker alarms; *dehydrated carcases*

Good housing

Comfort around resting

Dairy cattle	Time needed to lie down; collisions with housing equipment during lying down; animals lying outside lying area; cleanliness of the animals
Fattening cattle	Time needed to lie down; cleanliness of the animals
Veal calves	Lying positions; percentage of ruminating and lying animals; cleanliness (animals and environment)
Sows/piglets	Pressure injuries (sows); manure on body (sows)
Finishing pigs	Pressure injuries; manure on body
Laying hens	Plumage cleanliness; litter quality; useable area; type and state of flooring; characteristics of perches
Broilers	Plumage cleanliness; litter quality; floor area; atmospheric ammonia

Thermal comfort

Veal valves	Relative humidity; temperature; air flow
Sows/piglets	Shivering/ panting; social thermoregulation (huddling); environmental temperature
Finishing pigs	Shivering/panting; social thermoregulation (huddling); environmental temperature
Laying hens	Panting; social thermoregulation (huddling); ventilation; humidity; temperature alarms
Broilers	Panting; social thermoregulation (huddling); ventilation; humidity; temperature alarms

Ease of movement

Dairy cows	Presence of tethering; access to outdoor loafing area and/or pasture
Veal calves	Calves slipping when walking; slipperiness of the floor
Sows/piglets	Total pen space and stocking density; presence and size of stalls and/or farrowing crates (all sows)
Finishing pigs	Total pen space and stocking density
Broilers	Gait score; stocking density

Good health

Absence of injuries

Dairy cows	Lameness; integument alterations
Fattening cattle	Lameness; integument alterations
Veal calves	Skin alterations; claw and joint alterations; lameness; tail tip necrosis
Sows/piglets	Lameness; wounds on body
Finishing pigs	Lameness; wounds on body
Laying hens	Foot pad dermatitis; culls; predator protection; boundary fence effectiveness; cover on the range
Broilers	Foot pad dermatitis; hock burn; predator protection; cover on the range

Absence of disease

Dairy cows	Clinical scoring (e.g. nasal discharge, vulvar discharge); milk somatic cell count; mortality; culling rate
Fattening cattle	Clinical scoring (e.g. nasal discharge, bloated rumen); mortality; culling rate
Veal calves	Clinical scoring (e.g. increased respiratory rate, diarrhoea, anaemia); *pathological findings on lung, abomasum and rumen*
Sows/piglets	Clinical scoring (e.g. metritis, mastitis, scouring); health management; criteria for euthanasia; hygiene/ cleaning routine
Finishing pigs	Clinical scoring (e.g. twisted snouts, rectal prolapse, scouring); skin condition; apathy; health management; criteria for euthanasia; hygiene/ cleaning routine
Laying hens	Clinical scoring (e.g. comb/keel score, skin wounds, plumage cleanliness); mortality; culling rate; biosecurity measures; dust sheet test; inspection routines; disease and treatment records; time spent by stockman inspecting birds; hospitalisation of birds
Broilers	Clinical scoring (e.g. eye pathologies, enlarged crop); mortality; culling rate; *pathological findings (e.g. ascites, emaciation, pericarditis);* biosecurity measures; dust sheet test; inspection routines; disease and treatment records; time spent by stockman inspecting birds

Absence of pain induced by management procedures

Dairy cows	Procedure, age, use of anaesthetics/analgesics for dehorning and tail docking
Fattening cattle	Procedure, age, use of anaesthetics/analgesics for dehorning, tail docking and castration
Veal calves	Tail docking
Sows/piglets	Nose ringing; tail docking (sows) castration; tail docking; teeth clipping (piglets)
Finishing pigs	Castration; tail docking
Laying hens	Beak trimming severity; beak shape
Broilers	Effectiveness of perimeter fencing

Appropriate behaviour

Expression of social behaviours

Dairy cows	Agonistic behaviours
Fattening cattle	Agonistic behaviours; cohesive behaviours
Veal calves	Social horning, mounting, social licking
Sows/piglets	Positive social behaviour; negative social behaviour (all sows)
Finishing pigs	Positive social behaviour; negative social behaviour
Laying hens	Huddling; enrichment measures; aggressive behaviours; possibility for birds to escape aggressive behaviours
Broilers	Huddling; enrichment measures; aggressive behaviours

Expression of other behaviours

Veal calves	Play behaviour; maintenance behaviours
Sows/piglets	Environmental enrichment stereotypies (sows); exploratory behaviour
Finishing pigs	Environmental enrichment; exploratory behaviour
Laying hens	Natural light; spectral and flicker frequency of light; enrichment measures; characteristics and use of nests
Broilers	Natural light; enrichment measures; cover on the range

Good human-animal relationship

Dairy cattle	Avoidance distance at the feeding place; avoidance distance in the home pen
Fattening cattle	Avoidance distance at the feeding place
Veal valves	Reaction to the presence of humans (approach and touch)
Sows/piglets	Fear of humans (sows)
Finishing pigs	Fear of humans
Laying hens	Touch test; avoidance distance test; time spent by stockman inspecting birds; husbandry test; stockman interaction
Broilers	Touch test; avoidance distance test; time spent by stockman inspecting birds

Positive emotional state

Dairy cattle	Qualitative behaviour assessment
Fattening cattle	Qualitative behaviour assessment
Veal valves	Novel object test
Sows/piglets	Qualitative behaviour assessment
Finishing pigs	Qualitative behaviour assessment
Laying hens	Novel object test; qualitative behaviour assessment
Broilers	Novel object test; qualitative behaviour assessment

Table 6.3. List of measures in the operational protocols for the assessment of slaughterhouses for different species and animal categories that underwent practical application and were then further refined.

Good feeding

Absence of prolonged hunger
    Broilers   Feed withdrawal and journey times
Absence of prolonged thirst
    Finishing pigs   Water supply
    Broilers   Dehydrated carcases; water withdrawal and journey times
Good housing
Comfort around resting
    Finishing pigs   Stocking density and flooring of lorries; stocking density and flooring of lairage pens
Thermal comfort
    Finishing pigs   Shivering/panting; social thermoregulation (huddling); environmental temperature
    Broilers   Dead on arrival; panting in lairage
Ease of movement
    Fattening cattle   Slipping and falling (during unloading and driving to the stunning box)
    Finishing pigs   Slipping and falling during unloading
    Broilers   Stocking density in transport crates

Good health

Absence of injuries
    Fattening cattle   Carcase bruising
    Finishing pigs   Lameness; skin lesions
    Broilers   Skin damage; foot and toe damage; limb fractures; comb wounds; feather damage
Absence of disease
    Finishing pigs   Sick animals on arrival; dead animals on arrival; pathological findings (lung, pericardium, liver)
Absence of pain induced by management procedures
    Fattening cattle   Stunning effectiveness (eye movements, righting reflex, excessive kicking)
    Finishing pigs   Stunning effectiveness (corneal reflex, righting reflex, rhythmic breathing, vocalisations)
    Broilers   Birds flapping on the shackle line; birds receiving pre-stun shocks; birds not effectively stunned

Appropriate behaviour

Good human-animal relationship
    Finishing pigs   Vocalisations when driven to stunning area
Positive emotional state
    Fattining cattle   Fear behaviours (moving backwards, freezing, running, vocalisations) indicating negative emotional state
    Finishing pigs   Fear behaviours (reluctance to move, turning back) indicating negative emotional state

6.2.4 Need for clear descriptions of measures and protocols

The goal was (and still is) to produce protocols that can be used by third parties such as certification and inspection bodies, enforcement agencies, advisory bodies and research groups. Therefore, the standardised description and publication of the protocols is critical in order to ensure the proper application of the welfare measures and the subsequent calculation of overall welfare scores (see below). Furthermore, the guidelines for the certification decisions required for the product information system must be defined. These requirements can be met by the production of clear summary descriptions for each measure and for the integration systems (calculation of scores for the overall assessment of welfare).

Thus, all the standardised assessment protocols developed by Welfare Quality® follow the same general outline:
- title of the measure;
- scope: type of measure (i.e. animal-based or resource-/management-based) and animal category to which the measure applies;
- sample size;
- method (i.e. how to collect data on farm or at the slaughterhouse, including the scale of measurement at individual or group level);
- classification (i.e. the unit which is used for the calculation of welfare scores).

Additional information on sampling and further practicalities of the assessment (e.g. order in which the measures should be carried out during the farm visit) is provided. Two examples are shown in Box 6.3 and Box 6.4.

6.3 How the protocol was developed

6.3.1 Construction of the protocol from the measures

At this stage each of the numerous measures had been studied for its independent validity, reliability and feasibility as an indicator of welfare. The measures that met the requirements were combined and integrated into welfare assessment protocols that had a common approach across the species (cattle, poultry and pigs). One or more measures were chosen for each of the 12 welfare criteria (see Chapter 5 in order to yield an estimation of the overall level of animal welfare.

Practical (including economic) feasibility dictates that the animal welfare assessment system must be as concise, time-effective and easy to carry out as possible (van Reenen and Engel, 2009). Therefore, the feasibility of the welfare assessment system(s) was examined on a variety of commercial farms in order to evaluate the information

---

Box 6.3. Description of welfare measures: foot pad dermatitis in broilers.

Title	Foot pad dermatitis
Scope	Animal-based measure: Broiler chicken
Sample size	10 birds from each of 10 locations
Method description	Foot pad dermatitis is a contact dermatitis found on the skin of the foot, most commonly on the central pad, but sometimes also on the toes. The skin is turned dark by contact with litter and consequently deep skin lesions can result. The scoring scale allows an assessment of the severity of these lesions (see photographic reference). Assess the presence of hock burns with regard to the severity scale, scoring categories 0/1/2/3/4 as photographic illustration. Assess the number of animals in each scoring category and combine the categories for classification.
Classification	Individual level:
	a. No evidence of foot pad dermatitis (Score '0')
	b. Minimal evidence of foot pad dermatitis (Scores '1' and '2')
	c. Evidence of foot pad dermatitis (Scores '3' and '4')

| 0 | 1 | 2 | 3 | 4 |

© A. Butterworth, University of Bristol

---

provided by each welfare indicator in relation to all the other potential indicators. Further analysis of the findings enabled the refinement and ultimate definition of final assessment protocols. This approach consisted of three parts:

1. An analysis of the correlations and associations between different animal-based measures, between animal-based and resource-/environment-based measures, and the distributions and scales of various measurements. It was recognised that the welfare implications of some indicators might overlap and that their marginal value could be low so we considered whether such indicators could replace one another in the welfare assessment (i.e. if one indicator could be used rather than 2 or more to evaluate a particular criterion) or whether their combination strengthened the reliability of the assessment (Temple *et al.*, 2011).

Box 6.4. Description of welfare measures: time needed to lie down in dairy cows.

Title	Time needed to lie down
Scope	Animal-based measure: dairy cows
Sample size	Minimum of 6 lying down movements
Method description	This measure applies to lactating cows as well as to dry cows and pregnant heifers if they are kept with lactating animals. It considers all observable lying down movements (minimum sample size of 6 is required).
	Time recording of a lying down sequence starts when one carpal joint of the animal is bent and lowered (before touching the ground). The whole lying down movement ends when the hind quarter of the animal has fallen down and the animal has pulled the front leg out from underneath the body.
	Time needed to lie down is recorded in seconds, continuously in the focus segment. The duration of a lying down movement is only taken when undisturbed by other animals or human interaction and – in case of cubicles and littered systems – if it takes place on the supposed lying area. Observations take place in segments of the barn.
Classification	Individual level:
	Time in seconds
	Herd level:
	Mean time to lie down (in seconds)

---

2. A calibration of simplified versions of the monitoring system against the full version was carried out. Mainly for feasibility reasons, a limited number of measures that had been included in the original protocols were dropped from the lists (e.g. during human approach testing in dairy cattle, avoidance distance in the home pen was eliminated because of safety reasons and time constraints whereas avoidance distance at the feeding rack was retained).

3. A risk factor analysis was performed to identify potential risk factors for animal welfare; the resultant information was thought likely to improve the understanding of possible causes of reduced welfare in practice and to suggest ways of improving farm animal welfare. For example, if poor body condition has been identified as a major problem on a particular farm, resource based indicators (for example, quality and quantity of food provided to the animals) will be needed to identify the cause of the problem and the best strategies to solve it.

A farm or slaughterhouse visit, comprising different stages, was then designed.

6.3.2 Organisation of the farm or slaughter plant visit

The welfare assessment is carried out in accordance with the published protocols (Welfare Quality®, 2009a,b,c). When organising the visit basic information such as the timing of farm (milking, feeding, etc.) and slaughterhouse routines (unloading of trucks, etc.) is needed in advance. This information is usually obtained during the first telephone contact with the manager of the animal unit.

The first step of the visit involves interviewing the manager. Since the manager may not be familiar with the Welfare Quality® assessment the assessor begins by giving him/her an overview of the assessment protocol and of what is going to be done during the visit. General information about the farm or slaughterhouse is also recorded at that time by means of a questionnaire. Information on management, prevention of disease, feeding, hygiene, temperature regulation, castration routine, euthanasia criteria, and production and mortality records is collected. The manager is then asked to accompany the assessor during a walk around the building for visual inspection and sketching of the building. This enables the assessor to form an overview of the farm/slaughterhouse that can guide the choice of sampling strategy (e.g. how many rooms, ages, etc. should be tested). If the manager is busy, parts of the questionnaire can be completed at the end of the visit rather than the beginning. A clear benefit of completing the questionnaire at the end of the visit became apparent – the assessor gained a better understanding of the farmer's viewpoints and comments. In the slaughterhouse protocol, the questionnaire enables the gathering of information about the unloading and waiting areas for trucks, the emergency pens, stunning system, the use (or not) of electric prods when moving the animals, the presence or absence of the vet during unloading and the use of showers during lairage.

In the second stage, the assessment involves collecting data on the animals and the resources. A specific order in which the measures to be taken on farms or at slaughter is provided in the protocol for each animal type. In general, the animal-based assessment starts with measures recorded from outside the pen and by observing the whole group. In sows and growing pigs for example, the measures recorded from outside the pen consist of those related to the positive emotional state criterion (by means of the Qualitative Behavioural Assessment, QBA), the expression of social and other behaviours (scan sampling), and the presence of stereotypies, respiratory problems (coughing and sneezing), and thermal comfort measures (shivering, panting, huddling). Afterwards, the assessor enters the pen to assess the human-animal relationship and other animal-based measures related to the welfare principles of good feeding, housing, and health. Animals are individually scored for body condition, bursitis, shoulder sores, dirtiness (or presence of manure on the body), wounds on the body, tail biting, vulva lesions, lameness, pumping (heavy and laboured breathing), twisted snouts, rectal prolapse, uterine prolapse, skin condition, constipation, scouring, metritis, mastitis, local infections, tremor, splay leg and hernias. These measures are taken in approximately 30 pregnant sows, in 10 lactating sows and their litters, and/or in 150 growing pigs from 10 different pens. Some measures will require sampling of animals at specific stages of pregnancy (early, mid and late gestation) or at different stages of the growing/fattening period (at the beginning of the period but at least one week after mixing to avoid effects of the hierarchy formation, and at the end when space allowance is lower). Rather than entering many different pens and selecting lots of different animals the same animals should be used for as many different measures as possible to save time and minimise disturbance. The stage of pregnancy or growth is not considered likely to affect other measures. However, ensuring a representative sample simplifies the selection process. On many farms, animals in different stages may be housed within the same building (or even room), and are likely to be distributed equally across the building/room. However, if there are many small pens within a building or room those at either end of the building (and in the middle if necessary) should be selected. On farms where animals at the same biological stage are housed in different buildings it is important to sample animals from all the different types of buildings. Table 6.4 shows the order of the welfare assessment in dairy cows on farm.

The protocols at slaughterhouses include assessment of animal-, resource- and management-based measures (Dalmau *et al.*, 2009). The pig welfare assessment starts in the unloading area, where general fear, thermoregulatory behaviour, slipping and falling, lameness, sickness and mortality at arrival are measured. In lairage, eight pens are selected according to the time of arrival of animals, the size of the plant, and the distance to the stunning system. Behavioural thermoregulation measures, such as huddling, shivering or panting are scored, and the stocking density and the number of drinking points are also recorded here. When animals are moved from lairage to

Table 6.4. Order of welfare measures on a dairy farm.

Measure	Method
Avoidance distance	Observation
Qualitative behaviour assessment	Observation
Time to lie down	Observation
Social encounters	Observation
Clinical scoring	Observation
Resources	Check-list
Management	Questionnaire

the stunning area, the human-animal relationship is assessed at group level in terms of high-pitched vocalisations (squealing or screaming). Stunning effectiveness is assessed by the absence of corneal reflex, rhythmic breathing, righting reflex and vocalisations in 60 pigs divided into three batches of 20. After slaughter, the carcasses are examined for skin lesions, pleurisy and pneumonia in the lungs, pericarditis in the heart and white spots in the liver.

6.4 Practical testing of the protocols

One of the main objectives of the welfare assessment protocols is that they should be applicable to most farming conditions. This was evaluated through surveys on representative samples of different rearing systems and local conditions (e.g. climate) found in Europe and Latin America. Farms were selected for the survey on the basis of management practices, farm size, and veterinary records.

The protocols were tested on 696 farms (91 for dairy cows, 85 for beef cattle, 224 for veal calves, 90 for sows, 71 for growing pigs, 77 for laying hens and 58 for broilers) in 10 different countries. The dairy farms differed with respect to husbandry aspects such as the type of housing (loose or tethered), the presence of an outdoor 'loafing area', or access to pasture for cows. On some beef producing farms, the cattle were grazed whereas on others they were kept in intensive systems indoors. There were also notable differences between farms in the use of litter (straw). The sample of veal calf farms included those producing both so-called 'white veal' and 'pink veal'. These farms also varied according to the type and origin of calves, group size, size of the farm, diet (amount of milk replacer and amount and type of solid feed), climate control, daylight intensity, and management. The sow farms examined used extensive outdoors or indoor husbandry systems with different types of floor, bedding material, and feeding system. Like sow farms, the sample of farms with finisher pigs included outdoor and indoor units as well as ones that did or did not provide bedding material. Nine slaughterhouses were also visited to assess the welfare status of finishing pigs before (while being unloaded) and during slaughtering (stunning). The laying hen farms visited included free range, varying indoor floor systems, and those with conventional cages, furnished cages, and aviaries. On a limited number of broiler farms the birds had the opportunity to go outside.

It was concluded that many animal-based measures in the Welfare Quality® assessment protocols are sufficiently sensitive to show high variability and thereby allow discrimination between the farms. This aspect was more closely studied in pigs (Temple *et al.*, 2011).

6.4.1 Acceptability of the protocols by producers

In general, the protocols for all three species seem to work well, they are feasible and they received largely positive responses. More detailed information on the farmers' views was gathered from 63 beef farmers who were interviewed at the end of an assessment visit. Most were pleasantly surprised that the protocol involves little input on their part in data collection (thus not taking up their time) and that none of the measures are invasive or involve moving animals in or out of pens. Perhaps even more important is the high level of interest the farmers showed in the animal-based parameters; they are not usually provided with this sort of information. They were especially interested in the behavioural measures.

Some comments were made regarding validity, reliability or feasibility aspects of individual measures and on the assessment procedure in general. Some farmers pointed out that it would be better if bulls avoid an approaching person to a certain extent for two different reasons: (1) bulls should show respect for the farmer so that they can be safely handled; (2) bulls should have learned to retreat from the feeding rack so that the farmer can safely enter the pen to replace/replenish the litter.

Only a few farmers (6 of 63, 9.5%) disliked aspects of the assessment. These particular aspects were: economic components were not covered, tail-docked animals were recorded, the assessment did not consider that animals are normally dirtier in summer, the animals appeared nervous during the assessment and the farmers felt they did not know enough about the assessments.

Seven of the 63 farmers interviewed (11.3%) suggested improving the assessment by: filming some of the observed tests/measures to give the farmer an overview of which behaviours occur, carrying out observations once a week to once a month instead of three times a year, including economic data, assessing the feeding and taking account of the breed.

Most of the farmers (71%) did not propose additional measures but the others would include: feeding behaviour, feed quality and feed storage, behaviour at the drinker behaviour during the night (by video recordings), assessment of calves, barn climate, animal weight gain (to judge performance), use of medicines, breed, and system characteristics of the barn.

To summarise, farmers were quite satisfied with the protocol. Some of the additional measures they suggested are not feasible for the reasons discussed earlier, but others could be taken into consideration in revised versions of the protocols. These include for example the use of specific vocalisations (Moura *et al.*, 2008) or more frequent use of automated measurements, e.g. lying position as an indicator of thermal comfort (Shao and Xin, 2008). Of course, any new measure would need to be tested for validity, reliability and feasibility.

6.4.2 Feasibility of the protocol (duration, other factors)

As stated previously, for the overall protocol to be feasible it should require little input from the farmers, should be easy to perform in commercial situations, and should not take too much time. Traditionally, farm visits for the purpose of certification last about 2 h, which is less than needed for the Welfare Quality® assessment in its present form. In 2 h, an assessor can only take resource/management based measures, either directly on the farm (e.g. looking at housing conditions) or by interviewing the farmer (e.g. on the feeding, husbandry practices, etc.). This approach differs a great deal from a Welfare Quality® assessment where animal-based measures are not only considered the most effective ones for assessing welfare relevant to the principles and criteria identified by Welfare Quality® but also allow comparison of many different types of systems. However, this means that the additional time needed to carry out a full Welfare Quality® assessment (see below) could compromise its widespread acceptability.

The mean time taken to perform the full protocol in growing pigs at farm is 6 h and 20 min (±51 min) per visit (Table 6.5), ranging from 315 to 580 min (Temple *et al.*, 2011). The interview, the only part of the protocol that requires farmer participation, takes approximately 40 min (ranging from 25 to 60 min) depending on the farmer's level of interest. Furthermore, the time taken to record general information varies according to the size of the farm (as much as 200 min on the largest farm). Of course, obtaining a general plan of the farm before a visit could shorten this part of the protocol. Farm size and the distance between buildings also affected the time

Table 6.5. Time needed (min) to record the different parts of the Welfare Quality® protocol in growing pigs on farm (Temple et al., 2011).

Parts of the protocol	Mean	Standard deviation	Median	Min	Max
General information	72	33.0	60	30	200
Questionnaire alone	37	10.3	35	25	60
Qualitative behaviour assessment	30	5.5	30	20	45
Scan sampling of behaviours	86	14.7	85	70	120
Feeding, housing, health measures	130	22.9	125	100	170

required for QBA, scan sampling of behaviours, and the recording of good feeding, housing, and health measures. The numbers of animals per pen, the stocking density, the animals' behaviour (frightened or not), the dirtiness of pigs, and the light intensity within the buildings are other factors that might influence the time needed to record feeding, housing, and health measures.

In pig abattoirs the time required to collect the data is 3 h during unloading, 40 min in the lairage, 20 min from lairage to stunning, 30 min in the stunning area and 30-60 min after slaughter, with a mean total of 5 to 5.5 h (Dalmau *et al.*, 2009). Two of the most influential factors are: the frequency of truck arrivals and the time between the unloading and the beginning of slaughtering.

Completion of the welfare assessment takes between 5 to 8 h for dairy and beef farms, with the duration depending on herd size and the number of buildings (Table 6.6). Duration also depends on the ease and speed with which the assessor can measure certain parameters like behaviour (avoidance distance towards humans) or clinical indicators (e.g. skin lesions, signs of clinical disease). The experience and skills of the assessors also play a role. In very large herds, sample size and identification of representative samples may also become a problem. Currently, about 8 h is needed to carry out the full assessment in veal calf and laying hen units whereas it can be done in 4 h at broiler farms.

During the first year of the Welfare Quality® project several stakeholders (farmer organisations, breeding companies, etc.) expressed concern that running the Welfare Quality® assessment system would take too long and thus be too costly. This concern was made particularly clear at the second Welfare Quality® stakeholder conference (Berlin, May 3-4, 2007). However, the duration of farm visits did not worry the farmers that had been involved in the assessment exercise. For example, in the beef cattle implementation study 8 h was mentioned as an acceptable duration of farm

Table 6.6. Duration of farm visits according to the final Welfare Quality® protocols.

Animal type	Herd size (animals observed)	Duration (h)
Cows	50 (33)	5
Bulls	50 (33)	4
Pigs	(120-150)	6
Sows/piglets	(120-180)	5
Broilers	(100-150)	4
Layers	(100)	8

visits by external assessors and farmers. On average, farmers were willing to be present for 1.5 h.

The Welfare Quality® project investigated a number of alternative approaches in order to refine the form and duration of the assessment so that it became more acceptable to stakeholders while still ensuring reliable results. The options involved modification of either the completeness of the farm visit or of the intervals between farm visits. Reducing the duration of a farm visit in which all measures were still included could conceivably be achieved by reducing the sample size but statistical analyses showed that this would compromise the likelihood of obtaining a reliable picture. We also wondered if some animal-based measures could be replaced by management- and/ or resource-based ones or if we could assess fewer criteria. However, this approach is incompatible with the holistic nature of the assessment and the view that animal-based measures are more relevant and should thereby be prioritised over resource-based ones. The strength of the integrated approach lies in the use of the entire assessment protocol (see Chapter 7). The researchers strongly believe that the entire protocol should remain more or less as it is, though some refinement could occur when new knowledge or new techniques (e.g. automation) become available. Nevertheless, for the sake of practicality it may not always be necessary to collect data on every criterion at each visit. A possible solution could be to run an extensive assessment on the first visit but to then only measure particular features on subsequent visits. This would certainly reduce the workload. The chosen measures could correspond to weak points identified during the first visit and/or ones chosen at random. In that event the overall scoring of the farm would be based on the observed results and on the most recent data collected on a farm for each 'missing' measure (e.g. from the previous visit or an even earlier one). It was also proposed that the assessor could focus on the major welfare problems in a given population and just apply the measures and criteria related to these problems; it would thus be possible to check if slaughterhouses or farms comply with the requirements of specific criteria (e.g. non-compliance with the criterion of 'absence of prolonged hunger' if many lean animals are found) but no overall assessment would be produced (see Chapter 9).

The availability of protocols that can be used to assess welfare at farms using different production systems is important not only for reasons of harmonisation/ standardisation but also because this exercise would 'stress-test' the sensitivity of a particular protocol and help evaluate the feasibility of the overall protocol, especially in extensive conditions. The protocols that have been developed so far refer primarily to more or less intensive housing conditions. More extensive systems such as pasture-based milk production or beef production (e.g. cow-calf herds) have not been explicitly covered. In accordance with experience gained in the Latin-American subproject, some parts of the protocol might not be feasible in outdoor production systems for several reasons:

- It is difficult to observe some behaviours in large fields especially if the animals are widely spread, if they are situated on elevated areas that obstruct the assessor's view, or if the animals are in the shelters.
- Close observation of the animals, i.e. from a short distance, can disturb them and cause an alarm response. Therefore, as most of the health measures must be taken when the assessor is less than 2 meters from the animal (i.e. clinical scoring), visits need to be planned with the farmer to ensure that health measures can be taken when animals are grouped in the feeding area and behavioural measures taken some time before or after feeding time.
- In outdoor systems animals can modify their behaviour according to the weather conditions, e.g. less activity during heavy rain or high winds.
- While the behavioural parameters had been validated in intensive farms some, such as wallowing in pigs, still need to be validated in extensive conditions. Another example is the need to determine whether foraging should be considered as a measure of exploratory or of feeding behaviour.

On the other hand the way that poultry are kept is rather similar throughout the world, with the exception of very small (backyard) flocks and no major problems were experienced because of variations in housing systems or the weather. Even quite extensive free range birds are usually in the building during the night.

In conclusion, although some adaptations of the protocol may be necessary, it seems that it can be applied in a variety of farming systems and conditions.

6.4.3 Training

Clearly, a critical 'component' of all assessment systems is the assessor. Without competent and credible assessors, no certification scheme can function in a way that will satisfy both the producers and the consumers (Butterworth, 2009). The Welfare Quality® assessment protocols for cattle, pigs and poultry have been published and are easily available but these alone do not guarantee the capability to carry out the assessment; meaningful data can only be obtained when observers in each country have the same level of training. To standardise the implementation of the assessment protocol and achieve high repeatability between assessors, they must be continuously assessed during a robust training course until they develop a uniform scoring. When training professionals for the assessment of welfare on farm, some methodological aspects should be taken into account. Firstly, it is important that the professionals realise that an animal's welfare embraces its physical and mental state and that good animal welfare implies both fitness and a sense of well-being. The awareness that welfare is multidimensional resulted in the decision to base the Welfare Quality® assessment system on the four main principles of good animal welfare developed in the project. As the assessment relies largely on animal-based measures, the assessors should have a good understanding of the basic biology, physiology, health and behaviour of animals and the mechanisms they use to cope with difficult environments. The assessment will also involve recognition of symptoms of diseases, lesions, and signs of pain. As the assessors might not be veterinarians, the training does not aim to identify, diagnose or treat these problems, but rather to highlight the presence of injuries and health problems affecting the welfare of animals.

The assessors have to be fully trained in scoring the different measures firstly through classroom presentations and exercises, using photographs and video clips, and then in practical field exercises on farm or at the slaughterhouse (Velarde et al., 2010). During the classroom activity the rationales and execution of the measures are presented. Any doubts about the scoring of the different measures are discussed with the help of video clips and photographs designed to train the assessors. Later, the assessors are tested using photographs and video clips that have been previously scored by experts whose scores are used as gold standards for comparison with those given by the trainees. When the correlation between the gold standard and the assessor reaches a certain threshold, that assessor is considered fully trained for this measure. If the minimum correlation is not achieved, the assessors are re-trained by discussing some of the images scored previously according to the gold standard scores. An important consideration in any attempt to increase the repeatability and reliability of the assessment is that the measures should be simple to collect, in a way that minimises the need for value judgment.

The farm/slaughterhouse visit during training has several objectives. The first is to describe 'ad hoc' the assessment of the measures. The teacher demonstrates the procedures by assessing some animals and then asks the trainee to do the same. Finally, both assess some groups simultaneously and the results are compared. The second objective is to discuss those measures that had low correlations during the classroom sessions and that can be easily trained on farm/slaughterhouse. The third is to explain the sampling procedure, the order in which the measures are taken and practical aspects to take into account when approaching animals for different purposes. When returning to the classroom, at the end of the course, the trainee is asked to explain step by step how the welfare assessment procedure is carried out. The trainee inspectors are assessed several times during the training course to ensure that they develop a uniform scoring that retains objectivity, impartiality and repeatability. Later, when assessors have been active in the field for a set time period (to be determined), they should be 're-assessed' to ensure their maintenance of high quality assessment.

6.5 Conclusion: what lessons did we learn?

Much has been achieved during the process of developing the Welfare Quality® assessment system. Considerable effort was devoted to identifying appropriate measures for each criterion. Many important decisions were taken during the process of selecting measures that are valid, reliable and feasible. A science-based approach was used but it is also clear that ethical and pragmatic concerns played important roles. The experiments and literature studies that underpin the welfare assessment protocols for cattle, pigs and poultry are gathered together in three separate reports in the Welfare Quality® Report Series (Forkman and Keeling, 2009a,b,c); these are available for general use. The protocols contain standard descriptions of the measures, they explain what data should be collected and in what way, they specify the appropriate sample size and the order in which the different measures should be carried out.

One of the main aims of Welfare Quality® was to build a welfare assessment system that used only animal-based measures. However, for some species and some criteria this proved not to be possible so substitute resource-based measures were selected. It has always been the intention that the Welfare Quality® protocols should be updated in the light of new knowledge. Welfare assessment is a rapidly expanding area of animal welfare science and even in the time since the publication of the protocols some new animal-based measures may have been developed that could be considered in addition to or instead of some of the measures currently included in the Welfare Quality® protocols. In addition, measures that were not included at the time of protocol preparation are now becoming increasingly feasible because of continuing developments in automatic data collection technology. For instance, an automatic lameness scoring system could perhaps be used on farms with access to the appropriate equipment (e.g. accelerators attached to the legs of cows, a pressure plate that a cow has to walk across) (Rushen *et al.*, 2012). However, it should be borne in mind that no matter how much these automatic recordings systems are reduced in price, their relatively high cost means they are unlikely to be implemented in some regions or farms.

The main problem though is what to do in very extensive farming systems, where it may be difficult to approach animals closely enough to carry out any sort of lameness scoring at all on them. For milking cows this problem might be overcome by taking the measure when they are gathered together for milking, but this is not the case for other extensively kept animals. Arranging the inspection visit for a day when the animals are scheduled to be caught for other treatments or for sorting may be possible for animals that only have access to a limited run or a paddock, like poultry or pigs, but it is less feasible for beef cattle kept on very large ranges for long periods of time.

Another issue was that the initial Welfare Quality® protocol was considered by many stakeholders to take too long to complete and was thereby unlikely to be readily implemented. Two major concerns were expressed. Firstly it was felt that the full assessment would not only take up too much of the farmer's time but also disrupt routine practices on the farm. Interestingly though, our experience of working with the farmers themselves suggests that they did not consider the time issue to be a major problem since their time was only really needed for the interview. Secondly there were fears that the costs of running the assessment would be too high. Clearly, it is necessary to reduce the duration of the assessment not only to minimise costs but also to reduce disruption. Of course, the need to keep the visit as short as possible placed major constraints on the types of measures that could be incorporated. For example, some behavioural observations were shortened or even omitted (even though it was recognised that they could yield valuable information relevant to the principle of 'appropriate behaviour'). Despite this possible shortcoming, a number of practicable options were identified, for example running a full assessment on the first visit and following this with shorter ones that just focus on identified problems.

An informal survey of former Welfare Quality® partners in 2012 revealed that the Welfare Quality® protocols are being used or have recently been used in 46 different studies, most notably in dairy cows (http://www.welfarequality.net/network/44579/7/0/40). Such widespread application of the protocols confirms their perceived usefulness. The results of the latter studies will also ultimately contribute to the further development and improvement of the protocols.

References

Albright, J.L. and Arave, C.W. (1997). The behaviour of cattle. Cab International, Oxon, UK. 306 pp.

Algers, B., Nordensten, L. and Zimmermann, P. (2009a). Measuring distress vocalisation in poultry. In: Keeling, L. (ed.) Assessment of animal welfare measures in layers and broilers, pp. 95-102.

Algers, B., Westin, R. and Raj, M. (2009b). Assessment of pre-stunning shock and stunning effectiveness. In: Keeling, L. (ed.), Assessment of animal welfare measures for layers and broilers. Cardiff University, Cardiff, UK, pp. 67-73.

Barnett, J.L., Hemsworth, P.H. and Jones, R.B. (1993). Behavioural responses of commercially-farmed laying hen, to humans: evidence of stimulus generalization. Applied Animal Behaviour Science, 37, 139-146.

Blokhuis, H.J., Jones, R.B., Geers, R., Miele, M. and Veissier, I. (2003). Measuring and monitoring animal welfare: Transparency in the food product quality chain. Animal Welfare, 12, 445-455.

Boissy, A., Manteuffel, G., Jensen, M.B., Moe, R.O., Spruijt, B., Keeling, L., Winckler, C., Forkman, B., Dimitrov, I., Langbein, J., Bakken, M., Veissier, I. and Aubert, A. (2007). Assessment of positive emotions in animals to improve their welfare. Physiology and Behavior, 92, 375-397.

Boivin, X., Lensink, J., Tallet, C. and Veissier, I. (2003). Stockmanship and farm animal welfare. Animal Welfare, 12, 479-492.

Botreau, R., Bonde, M., Butterworth, A., Perny, P., Bracke, M.B.M., Capdeville, J. and Veissier, I. (2007). Aggregation of measures to produce an overall assessment of animal welfare: Part 1 – A review of existing methods. Animal, 1, 1179-1187.

Bracke, M.B.M. and Spoolder, H.A.M. (2008). Novel object test can detect marginal differences in environmental enrichment in pigs. Applied Animal Behaviour Science, 109, 39-48.

Butterworth, A. (2009). Animal welfare indicators and their use in society. In: Smulders, J.F.M. and Algers, B. (eds.) Welfare of production animals: assessment and management of risks. Food Safety Assurance and Veterinary Public Health series no.5. Wageningen Academic Publishers, Wageningen, the Netherlands, pp. 371-390.

Courboulay, V., Meunier-Salaün, M.C., Edwards, S.A., Guy, J.H. and Scott, K. (2009). Repeatability of abnormal behaviour. In: Forkman, B. and Keeling, L. (eds.) Assessment of welfare measures for sows, piglets and fattening pig. Cardiff university, Cardiff, UK, pp. 131-140.

Dalmau, A., Geverink, N.A., Van Nuffel, A., Van Steenbergen, L., Van Reenen, K., Hautekiet, V., Vermeulen, K., Velarde, A. and Tuyttens, F.A.M. (2010). Repeatability of lameness, fear and slipping scores to assess animal welfare upon arrival in pig slaughterhouses. Animal, 4, 804-809.

Dalmau, A., Temple, D., Rodríguez, P., Llonch, P. and Velarde, A. (2009). Application of the Welfare Quality® protocol at pig slaughterhouses. Animal Welfare, 18, 497-505.

Engel, B., Bruin, G., Andre, G. and Buist, W. (2003). Assessment of observer performance in a subjective scoring system: Visual classification of the gait of cows. Journal of Agricultural Science, 140, 317-333.

European Commission (2007a). Attitudes of consummers towards the welfare of farmed animals – Wave 2. Special Eurobarometer 229(2)/Wave 64.4 – TNS Opinion & Social. http://ec.europa.eu/food/animal/welfare/survey/sp_barometer_fa_en.pdf. 60 pp.

European Commission (2007b). Attitudes of EU citizens towards Animal Welfare – Special eurobarometer 270/Wave 66.1 – TNS Opinion & Social. http://ec.europa.eu/food/animal/welfare/survey/sp_barometer_fa_en.pdf. 51 pp.

Fleiss, J.L., Levin, B. and Paik, M.C. (2003). The measurement of interrater agreement. In: Fleiss, J.L., Levin, B. and Paik, M.C. (eds.) Statistical methods for rates and proportions. John Wiles & Sons, Hoboken, NY, USA, pp. 598-626.

Flower, F.C. and Weary, D.M. (2006). Effect of hoof pathologies on subjective assessments of dairy cow gait. Journal of Dairy Science, 89, 139-146.

Forkman, B., Heiskanen, T., Graml, C. and Waiblinger, S. (2009). Assessment of general fearfulness. In: Forkman, B. and Keeling, L. (eds.), Assessment of animal welfare measures for layers and broilers, pp. 91-94.

Forkman, B. and Keeling, L. (eds.) (2009a). Assessment of animal welfare measures for dairy cattle, beef bulls and veal calves. Cardiff university, Cardiff, UK, 297 pp.

Forkman, B. and Keeling, L. (eds.) (2009b). Assessment of animal welfare measures for layers and broilers. Cardiff university, Cardiff, UK, 176 pp.

Forkman, B. and Keeling, L. (eds.) (2009c). Assessment of animal welfare measures for sows, piglets and fattening pigs. Cardiff university, Cardiff, UK. 310 pp.

Geverink, N.A., Meuleman, M., Van Nuffel, A., Van Steenbergen, L., Hautekiet, V., Vermeulen, K., Lammens, V., van Reenen, C.G. and Tuyttens, F.A.M. (2009). Repeatability of a lameness score measured on farm. In: Forkman, B. and Keeling, L. (eds.) Assessment of animal welfare measures for sows, piglets and fattening pigs. Cardiff university, Cardiff, UK, pp. 73-78.

Graml, C., Waiblinger, S. and Niebuhr, K. (2008). Validation of tests for on-farm assessment of the hen-human relationship in non-cage systems. Applied Animal Behaviour Science, 111, 301-310.

Graml, C., Waiblinger, S. and Niebuhr, K. (2009). Assessment of the human-animal relationship in laying hens. In: Forkman, B. and Keeling, L. (eds.) Assessment of animal welfare measures for layers and broilers, pp. 75-90.

Hemsworth, P.H., Barnett, J.L. and Jones, R.B. (1993). Situational factors that influence the level of fear of humans by laying hens. Applied Animal Behaviour Science, 36, 197-210.

Hemsworth, P.H. and Coleman, G.J. (eds.) (1998). Human-livestock interactions: the stockperson and the productivity and welfare of intensively farmed animals. Cab International, Oxon, UK, 152 pp.

Jones, R.B. (1987). Fear and fear responses: a hypothetical consideration. Medical Science Research, 15, 1287-1290.

Jones, R.B. (1997). Fear and distress. In: Appleby, M.C. and Hughes, B.O. (eds.) Animal welfare. CAB International, Oxon, UK, pp. 75-87.

Jones, R.B. and Boissy, A. (2011). Fear and other negative emotions. In: Appleby, M.C., Mench, J.A., Olsson, I.A.S. and Hughes, B.O. (eds.) Animal welfare. CAB International, Oxon, UK, pp. 78-97.

Knierim, U. and Winckler, C. (2009). On-farm welfare assessment in cattle: Validity, reliability and feasibility issues and future perspectives with special regard to the Welfare Quality® approach. Animal Welfare, 18, 451-458.

Knowles, T. G., Warriss, P.D., Brown, S.N., Edwards, J.E. and Mitchell, M.A. (1995). Response of broilers to deprivation of food and water for 24 hours. British Veterinary Journal, 151, 197-202.

Laister, S., Brörkens, N., Lolli, S., Zucca, D., Knierim, U., Minero, M., Canali, E. and Winckler, C. (2009a). Reliability of measures of agonistic behaviour in dairy and beef cattle. In: Forkman, B. and Keeling, L. (eds.) Assessment of animal welfare measures for dairy cattle, beef bulls and veal calves. Cardiff university, Cardiff, UK, pp. 95-112.

Laister, S., Brörkens, N., Minero, M., Lolli, S., Zucca, D., Knierim, U., Canali, E. and Winckler, C. (2009b). Reliability of measures of socio-positive and play behaviour in dairy and beef cattle. In: Forkman, B. and Keeling, L. (eds.) Assessment of animal welfare measures for dairy cattle, beef cattle and veal calves, pp. 175-188.

Laister, S., Stockinger, B., Regner, A.M., Zenger, K., Knierim, U. and Winckler, C. (2011). Social licking in dairy cattle-Effects on heart rate in performers and receivers. Applied Animal Behaviour Science, 130, 81-90.

Lensink, B.J., Van Reenen, C.G., Engel, B., Rodenburg, T.B. and Veissier, I. (2003). Repeatability and reliability of an approach test to determine calves' responsiveness to humans – a brief report. Applied Animal Behaviour Science, 83, 325-330.

Main, D.C.J., Whay, H.R., Leeb, C. and Webster, A.J.F. (2007). Formal animal-based welfare assessment in UK certification schemes. Animal Welfare, 16, 233-236.

March, S., Brinkmann, J. and Winkler, C. (2007). Effect of training on the inter-observer reliability of lameness scoring in dairy cattle. Animal Welfare, 16, 131-133.

Martin, P. and Bateson, P. (2007). Measuring behaviour. Cambridge University Press, Cambridge, UK. 176 pp.

Menke, C., Waiblinger, S., Fölsch, D.W. and Wiepkema, P.R. (1999). Social behaviour and injuries of horned cows in loose housing systems. Animal Welfare, 8, 243-258.

Moura, D.J., Silva, W.T., Naas, I.A., Tolón, Y.A., Lima, K.A.O. and Vale, M.M. (2008). Real time computer stress monitoring of piglets using vocalization analysis. Computers and Electronics in Agriculture, 64, 11-18.

Plesch, G., Broerkens, N., Laister, S., Winckler, C. and Knierim, U. (2010). Reliability and feasibility of selected measures concerning resting behaviour for the on-farm welfare assessment in dairy cows. Applied Animal Behaviour Science, 126, 19-26.

Polten, R. (ed.) (2007). Proceedings of the workshop 'Animal welfare improving by labelling?', 28 March 2007, Brussels, Belgium. 86 pp.

Pritchard, J.C., Barr, A.R.S. and Whay, H.R. (2006). Validity of a behavioural measure of heat stress and a skin tent test for dehydration in working horses and donkeys. Equine Veterinary Journal, 38, 433-438.

Raussi, S., Lensink, B.J., Boissy, A., Pyykkönen, M. and Veissier, I. (2003). The impact of social and human contacts on calves' behaviour and stress responses. Animal Welfare, 12, 191-203.

Regula, G., Danuser, J., Spycher, B. and Wechsler, B. (2004). Health and welfare of dairy cows in different husbandry systems in Switzerland. Preventive Veterinary Medicine, 66, 247-264.

Reinhardt, V. and Reinhardt, A. (1982). Mock fighting in cattle. Behaviour, 81, 1-13.

Rousing, T. and Waiblinger, S. (2004). Evaluation of on-farm methods for testing the human-animal relationship in dairy herds with cubicle loose housing systems – test-retest and inter-observer reliability and consistency to familiarity of test person. Applied Animal Behaviour Science, 85, 215-231.

Rushen, J., Chapina, N. and De Passillé, A.-M. (2012). Automated monitoring of behavioural-based animal welfare indicators. Animal Welfare, 21, 339-350.

Rushen, J., Pombourcq, E. and De Passille, A.M. (2007). Validation of two measures of lameness in dairy cows. Applied Animal Behaviour Science, 106, 173-177.

Salzen, E.A. (1979). The ontogeny of fear in animals. In: Sluckin, W. (ed.) Fear in animals and man. Van Nostrand Reinhold, New York, NY, USA, pp. 125-163.

Sato, S., Tarumizu, K. and Hatae, K. (1993). The influence of social factors on allogrooming in cows. Applied Animal Behaviour Science, 38, 235-244.

Schulze Westerath, H., Laister, S., Winckler, C. and Knierim, U. (2009). Exploration as an indicator of good welfare in beef bulls: An attempt to develop a test for on-farm assessment. Applied Animal Behaviour Science, 116, 126-133.

Scott, E.M., Nolan, A.M. and Fitzpatrick, J.L. (2001). Conceptual and methodological issues related to welfare assessment: a framework for measurement. Acta Agriculturae Scandinavica, Section A, Animal Science Supplementum, 30, 5-10.

Scott, K., Laws, D.M., Courboulay, V., Meunier-Salaun, M.-C. and Edwards, S.A. (2009). Comparison of methods to assess fear of humans in sows. Applied Animal Behaviour Science, 118, 36-41.

Shao, B. and Xin, H. (2008). A real-time computer vision assessment and control of thermal comfort for group-housed pigs. Computers and Electronics in Agriclture, 62 15-21.

Temple, D., Dalmau, A., Ruiz de la Torre, J., Manteca, X. and Velarde, A. (2011). Application of the Welfare Quality® protocol to assess growing pigs kept under intensive conditions in Spain. Journal of Veterinary Behavior, 6, 138-149.

Van Reenen, K. and Engel, B. (2009). Testing the assessment system: refinement and definition. In: Keeling, L. (ed.) An overview of the development of the Welfare Quality® project assessment systems. Cardiff university, Cardiff, UK, pp. 33-42.

Veissier, I., Butterworth, A., Bock, B. and Roe, E. (2008). European approaches to ensure good animal welfare. In: Rushen, J. (ed.) Farm animal welfare since the Brambell report. Elsevier, Amsterdam, the Netherlands, pp. 279-297.

Veissier, I., Capdeville, J. and Delval, E. (2004). Cubicle housing systems for cattle: comfort of dairy cows depends on cubicle adjustment. Journal of Animal Science, 82, 3321-3337.

Velarde, A., Dalmau, A. and Manteca, X. (2010). Animal welfare training for professional: moving from inputs to outcomes. 1st International Conference on Animal Welfare Education, Brussels, Beglium.

Waiblinger, S., Boivin, X., Pedersen, V., Tosi, M.V., Janczak, A.M., Visser, E.K. and Jones, R.B. (2006). Assessing the human-animal relationship in farmed species: A critical review. Applied Animal Behaviour Science, 101, 185-242.

Waiblinger, S., Knierim, U. and Winckler, C. (2001). The development of an epidemiologically based on-farm welfare assessment system for use with dairy cows. Acta Agriculturae Scandinavica – Section A: Animal Science suppl., 30, 73-77.

Waiblinger, S., Mülleder, C., Schmied, C. and Dembele, I. (2007). Assessing the animals' relationship to humans in tied dairy cows: Between-experimenter repeatability of measuring avoidance reactions. Animal Welfare, 16, 143-146.

Welfare Quality® (2009a). Welfare Quality® assessment protocol for cattle (fattening cattle, dairy cows, veal calves). Welfare Quality® Consortium, Lelystad, the Netherlands, 182 pp.

Welfare Quality® (2009b). Welfare Quality® assessment protocol for pigs (sows and piglets, growing and finishing pigs). Welfare Quality® Consortium, Lelystad, the Netherlands, 114 pp.

Welfare Quality® (2009c). Welfare Quality® assessment protocol for poultry (broilers, laying hens). Welfare Quality® Consortium, Lelystad, the Netherlands, 114 pp.

Wemelsfelder, F., Knierim, U., Schulze Westerath, H., Lentfer, T., Staack, M. and Sandilands, V. (2009a). Qualitative behaviour assessment. In: Keeling, L. (ed.) Assessment of animal welfare measures for layers and broilers, pp. 113-119.

Wemelsfelder, F. and Millard, F. (2009). Qualitative behaviour assessment. In: Keeling, L. (ed.) Assessment of animal welfare measures for sows, piglets and fattening pigs, pp. 213-219.

Wemelsfelder, F., Millard, F., De Rosa, G. and Napolitano, F. (2009b). Qualitative behaviour assessment. In: Keeling, L. (ed.) Asssessment of animal welfare measures dairy cattle, beef bulls and veal calves, pp. 215-224.

Westin, R., Velarde, A., Dalmau, A. and Algers, B. (2009). Assessment of stun quality in cattle. In: Keeling, L. (ed.) Assessment of animal welfare measures for dairy cattle, beef cattle and veal calves, pp. 89-93.

Winckler, C. and Willen, S. (2001). The reliability and repeatability of a lameness scoring system for use as an indicator of welfare in dairy cattle. Acta Agriculturae Scandinavica – Section A: Animal Science, 30, 103-107.

Chapter 7. Integration of data collected on farms or at slaughter to generate an overall assessment of animal welfare

Raphaëlle Botreau, Christoph Winckler, Antonio Velarde, Andy Butterworth, Antoni Dalmau, Linda Keeling and Isabelle Veissier

7.1 Introduction

One objective of the WelfareQuality® project was to propose a standardised assessment method that could be used to provide transparent information on farm animal welfare to all relevant stakeholders. As described in earlier chapters, Welfare Quality® therefore built welfare assessment systems for cattle, pigs and poultry incorporating numerous measures based preferably on the animals but also to a lesser extent on resources and management of animal units (farms or slaughter plants). Of course, the substantial amount of data gathered during assessment needs to be meaningfully interpreted in terms of welfare and then integrated to provide an overall evaluation of the animal unit. Therefore, Welfare Quality® designed a scoring model to integrate the data and to translate value judgements into refined and easily understandable information that could serve various purposes and guide the decisions of stakeholders (including consumers) with regard to farm animal welfare.

At the very start of the project it was determined that the assessment system developed must enable the differentiation of animals' living conditions in terms of their welfare. In other words, it should be sufficiently sensitive to identify and (once the data are integrated) to quantify variations between farms.

In this chapter, we explain the methods used to calculate welfare scores that express the overall level of animal welfare on the assessed farms or slaughter plants. We also describe some of the outcomes of running the Welfare Quality® assessment systems on approximately 700 farms across Europe.

7.2 The scoring system: how we interpret and synthesise the information

The various measures incorporated in the welfare assessment protocol(s) generate a very substantial amount of data which then needs to be integrated so that an overall evaluation of the animal unit can be given. To address this requirement and to facilitate the broad acceptance of the assessment system, Welfare Quality® designed a scoring model to integrate and ultimately transform all the data into a simple single welfare score.

Within the time constraints of the Welfare Quality® project the researchers managed to design scoring models for the welfare of dairy cows, fattening cattle, veal calves, fattening pigs, and broilers during the time they spend on farms. This effort largely excluded transport and slaughter except when measures taken at slaughter could be used retrospectively to assess the animals' welfare on the farm.

7.2.1 Constraints surrounding the development of scoring models

The first of a number of difficulties that arose reflected the multidimensional nature of animal welfare. Attempting to build an overall welfare assessment of a farm using the numerous measures taken on that farm can sometimes be considered as akin to adding things together that are as different as chalk and cheese. In other words, tests, measures and states that are often very different in nature have to be combined to form one meaningful whole. If this integration is not done appropriately there is a risk that some welfare problems might be hidden.

The second main challenge was that on the one hand concrete data are produced to describe farms (e.g. on farm A, 2% cows are very lean, 5% are lame, 50% cows flee when approached to 1 m, etc.) while on the other hand a judgement has then to be made from the consideration and interpretation of such data. And a judgement cannot be value free. The challenge this poses becomes even more critical when a judgement has to be made not just on the basis of a single measure (e.g. does a finding of 2% lean cows reflect good or bad welfare?) but also when it has to be formed according to a group of measures (e.g. do findings of 2% lean cows + 5% lame cows + 15% fearful cows, etc. reflect good or bad welfare?). Not surprisingly, ethical dilemmas arose when building the model for the overall assessment of the welfare of animals; these are described below:

1. The welfare of an animal is considered to be a matter of how that animal experiences its life (Bracke *et al.*, 1999). When a single aspect of welfare is considered, the animal's point of view can often be obtained, at least when short term preferences are considered, e.g. demand curves can reveal which of various foods is preferred (Holm *et al.*, 2007). However, it seems simply impossible to determine how an animal would itself perceive the severity of or attempt to rank very different aspects of welfare, e.g. being afraid of something versus being sick. The assessment of an animal's welfare is thus to some extent done from an expert's point of view. Hence, the result of a welfare assessment may depend on the assessors' experience and understanding. In that case, the question arises of who has the best expertise to estimate how an animal perceives it environment?
2. Already at the level of the individual animal, ethical decisions have to be taken as to how to integrate (or 'add-up') measures of the many different aspects of welfare into a single value. This can itself pose problems. First, for example some people

may consider certain aspects of welfare more important than other ones while other people may believe the opposite; a prime example is that some people may rank health higher than behavioural indices while others may consider behaviour to be more important (Fraser, 1995). Second, some people may allow compensation between different aspects of welfare while others may not.

3. A greater problem arises when one wishes to determine the welfare of groups of animals. Should we then concentrate on the animals that are in the poorest condition or should we consider the 'average' welfare status across all the animals in that group/herd/flock? In the latter case it is wholly unrealistic to 'ask' a group of animals to judge their overall welfare, for example by using demand curves. Once again, the assessment of the overall welfare status of a group of animals must inevitably be based on expert opinion.

4. When building an overall picture of welfare on a farm, it may be argued that the midpoint of possible assessment outcomes should corresponds to what is normally found in practice. This view allows the results obtained at a particular farm to be compared to those of other farms from the same population, i.e. a farm can be considered better or worse than another (Whay *et al.*, 2003). Conversely, it may also be argued that simply doing a little bit better than another farm or farms is not necessarily enough to achieve good animal welfare (Bekoff, 2008). According to such views, the assessment of welfare should be based on what is theoretically considered excellent, good, acceptable, etc. and not simply 'better or worse than'.

Clearly, ethical decisions had to be reflected in the Welfare Quality® scoring system. The partners in charge of developing the scoring model did not take these decisions themselves. Instead, a flexible model was designed in order to most accurately represent the most probable ethical decisions. Therefore, the Welfare Quality® partners consulted a number of people with relevant knowledge inside and outside the project about this flexible model (see below). The model was subsequently refined according to the results of these consultations. A close collaboration between animal scientists and a mathematician specialising in multicriteria decision problems was an essential part of this process because it enabled us to take into account the various viewpoints of a range of stakeholders when designing the scoring model.

We also faced some technical difficulties; the main problem was that data could be collected on very different scales:

- Some data were expressed on a nominal scale, e.g. the method used to dehorn young cattle (cauterisation by hot iron or by caustic paste).
- Some data were expressed on an ordinal scale, i.e. as ordered categories. For example, piglets can have a normal weight, be lean, or even very lean. In that case one does not know if the difference between normal and lean is of the same magnitude as the difference between lean and very lean.

- Other data were expressed on cardinal scales, e.g. the distance at which an animal turns away or flees when approached by a human being.

Even in the case of data recorded on a cardinal scale, the interpretation may not be linear. For instance, it is likely that one would consider that a cow that flees when the human is 3.5 m away does not differ meaningfully from a cow that flees at 3 m, but the same observer might ascribe very different values to a cow that flees at 0.5 m and to one that can actually be touched by the observer (i.e. an approach distance of 0 m) even though the difference in distance is the same (0.5 m) in both these examples.

All observations should thus be recorded on a common value scale that expresses the level of welfare of the animals.

7.2.2 Architecture of the model

The Welfare Quality® principles and criteria of good welfare were formulated using a top-down approach, i.e. the principles were defined first, then the criteria, and finally appropriate measures were selected to check for compliance with the criteria (Chapter 5). In contrast, the achievement of an overall assessment of animal welfare on a particular animal unit follows a bottom-up approach (Figure 7.1). First, criterion-scores are calculated from the results of the various measures, then principle-scores are calculated and finally the overall assessment is produced.

7.2.3 Consultation process

The scoring model was developed using methods that allow ethical decisions to be taken where it is deemed necessary, e.g. allowing or not allowing compensation between principles or criteria; taking into account just the animals in the worst

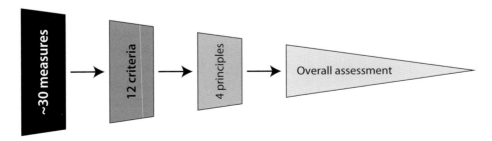

Figure 7.1. The bottom-up approach chosen in Welfare Quality® to produce an overall assessment of animal welfare.

condition or all animals in a group, etc. As described above, during the construction of the model, we identified and consulted a number of people (see below) from various backgrounds and then adjusted the model according to a considered synthesis of their opinions. The types of people consulted varied at each step in the process:

- The first step of the scoring model involves transforming the data obtained from the welfare measures into criterion-scores. Here we consulted the animal scientists from the Welfare Quality® project who had actually worked on and developed the measures. It was assumed that they were the people who knew the most about the meaning of these measures in terms of animal welfare. Furthermore, they already had experience of the kinds of results that can be found in practice. Between 4 and 8 scientists were consulted depending on the criterion and the animal species.
- Both animal and social scientists who had played leading roles in the Welfare Quality® project were consulted in the second step, where criterion-scores are converted to principle-scores. The social scientists brought in the points of view of the various stakeholder populations that they were studying (citizens, retailers, producers) while the animal scientists were considered to represent the animals' point of view. Between 13 and 18 scientists were consulted depending on the animal species.
- A mix of animal scientists, sociologists and stakeholders were consulted in the third step which involved converting principle-scores into an overall welfare assessment. The stakeholders were members of the Advisory Committee for the Welfare Quality® project; collectively this committee represented producer organisations, retailers, the food service sector, animal breeders, veterinarians, NGOs, certification bodies, the European Society for Agricultural Ethics, and other relevant institutions (e.g. a representative of the International Organisation for Animal Health was an observer). Stakeholders' opinions were considered essential at this stage because the production of a final overall welfare assessment very much depends on how the system is going to be used.

At no time were these people asked to make choices between ethical dilemmas. Datasets were shown to them and they were simply asked to react to these (examples are provided in Tables 7.1 and 7.2) by assigning values on a 0-100 scale where 0 corresponds to the worst situation one could expect to find on an animal unit and 100 the best one. In addition, it was established that a farm that obtains scores of 50 was to be considered 'not bad – not good' whereas a farm that scored below 20 ran a high risk of being excluded from any welfare scheme, i.e. deemed 'not acceptable'. A score of '20' was regarded as merely corresponding to legislative requirements or their equivalent.

For the final step of the scoring model scientists and stakeholders discussed the potential uses of the assessment system as well as the various ways of aggregating the four principle-scores while still taking into account the distribution of results across

Table 7.1. Dataset given to animal scientists to assign a score for absence of injuries according to the percentage of calves affected by moderate injuries (such as hairless spots) or severe injuries (such as joint lesions). Animal scientists were invited to rank the (virtual) farms from worst to best and then to assign a score (0-100) to each of them.

	% calves with no injury	% calves with moderate injuries	% calves with severe injuries	Rank	Score
Farm 1	100	0	0	?	?
Farm 2	0	100	0	?	?
Farm 3	0	0	100	?	?
Farm 4	50	50	0	?	?
Farm 5	0	50	50	?	?
Farm 6	50	0	50	?	?
Farm 7	50	25	25	?	?
Farm 8	20	50	30	?	?
Farm 9	0	25	75	?	?
Farm 10	75	25	0	?	?
Farm 11	75	0	25	?	?

Table 7.2. Combinations of criterion-scores given to animal and social scientists to assign principle-scores.

Score for criterion 'absence of prolonged hunger'	Score for criterion 'absence of prolonged thirst'	Score for principle 'good feeding'
25	75	?
40	60	?
50	50	?
60	40	?
75	25	?

farms visited by Welfare Quality® scientists. These discussions took place during face to face meetings between the two groups and by electronic correspondence.

Once the scoring model had been developed, it was also discussed in citizens' and farmers' juries in the United Kingdom, Italy, the Netherlands and Norway (Miele *et al.*, 2011).

7.2.4 Tuning of the model

From measures to welfare criteria

Because the total number of measures, the scale on which they are expressed, and the relative importance of measures differ, the calculation of scores varies between criteria and animal types. Three main ways of calculating criterion-scores are used:

- When all measures used to check a criterion are taken at farm level and are expressed in a limited number of categories, a decision tree is designed. An example is provided in Box 7.1.
- Whenever compliance with a criterion is checked by only one measure taken at individual level but with several degrees (e.g. moderate vs. severe problems), the proportion of animals observed can be calculated (e.g. percentage of animals walking normally, percentage of moderately lame animals, percentage of severely lame animals). In this case a weighted sum is calculated, with weights increasing with severity. An example is provided in Box 7.2.

Box 7.1. Decision tree as applied to absence of prolonged thirst in fattening pigs.

Thirst is not assessed directly on animals because signs of dehydration can be detected only in extreme cases. Instead, the number of drinking places, their functioning and their cleanliness are assessed. Adherence to the recommended number of pigs per drinker (10 pigs per functioning drinker and 5 for a drinker of reduced capacity) is established. If there are more pigs in the pen than recommended then the number of drinking places is considered insufficient. Thereafter, cleanliness of drinkers and whether or not pigs have access to two drinkers in the same pen is considered. The following decision tree is applied:

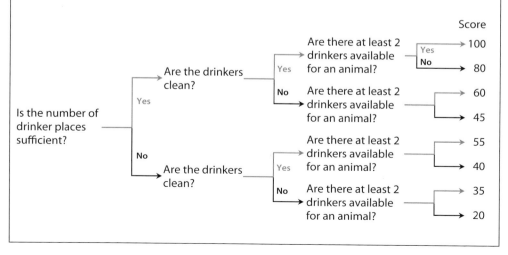

Box 7.2. Weighted sum and I-spline functions as applied to lameness in dairy cows.

The % of animals moderately lame and the % of animals severely lame are combined in a weighted sum, with a weight of 2 for moderate alterations and 7 for severe ones. This sum is then transformed into an index that varies from 0 to 100:

$$\text{Index for lameness I} = 100 - \frac{2(\%moderate) + 7(\%severe)}{7}$$

This index is computed into a score using I-spline functions:

When I ≤78, then Score = $(0.0750111 \times I) - (2.42066E^{-5} \times I^2) + (4.49587E^{-5} \times I^3)$
When I ≥78, then Score = $2129.52 + (81.9797 \times I) - (1.05008 \times I^2) + (0.0045324 \times I^3)$

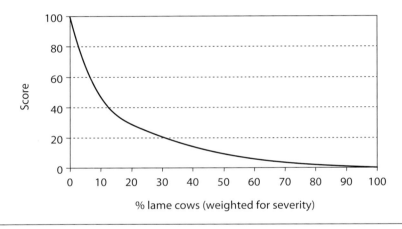

- When the measures used to check a criterion result in data that are expressed on different scales (e.g. percentage of animals lying outside the lying area, average latency to lie down, number of animals colliding while lying down), the actual data are compared to an 'alarm threshold' that represents the limit between what is considered to require action to safeguard welfare or not). Subsequently, the number of 'alarms' is valued. An example is provided in Box 7.3.

The animal scientists consulted were asked to interpret the raw data in terms of welfare (in this case they were scoring virtual farms). In the event that weighted sums had to be calculated, this consultation was used to define weights that produce the same ranking of farms as the one given by experts. The scores the scientists assigned were then used to define functions to transform data into scores. This exercise revealed that experts did not follow a linear reasoning, e.g. for a given disorder a 10% increase did not yield the same decrement in scores obtained at the bottom of the 0-100 scale (where most animals show this disorder) and at the top of the scale (when most

---

Box 7.3. Alarm thresholds applied to absence of diseases in broilers.

In broiler chickens the following disorders are checked on the farm or at slaughter: ascitis, dehydration, septicaemia, hepatitis, pericarditis, subcutaneous abscesses, mortality and culling. The incidence of each disorder is compared to an alarm threshold, defined as the incidence above which a health plan is required at the farm level.

Disorder	Alarm threshold (%)
Ascitis	1
Dehydration	1
Septicaemia	1.5
Hepatitis	1.5
Pericarditis	1.5
Subcutaneous abscess	1
Mortality	
< 20% due to culling	6
20% – 50% due to culling	7
≥ 50% due to culling	8

When the incidence observed on a farm reaches half the alarm threshold, a warning is given. The number of alarms and warnings detected on a farm are calculated. They are used to calculate a weighted sum finally transformed into a score using I-spline functions (as in the example shown in Box 7.2).

---

animals are normal/not affected). It is therefore necessary to resort to non-linear functions to produce criterion scores, in this case I-spline functions were used. Briefly, I-spline functions allow calculation of portions of curves so as to obtain a resulting smooth monotonic curve. They are expressed in the form of cubic functions (Box 7.2).

When a criterion was composed of very different items that experts found difficult to consider together at first glance, blocks of measures were first considered and these were then further aggregated using Choquet integrals (see below).

Although it is not generally the case, some measures may be related to several criteria (e.g. low body condition score can originate from hunger or disease, or both). In order to avoid double counting, measures were allocated to only one criterion, except in the very few cases where the measures were interpreted differently in one criterion than in another (e.g. access of cattle to pasture is used to check the ease of movement

– especially for animals which are tethered in winter – as well as the expression of behaviour).

From welfare criteria to welfare principles

Once the criterion-scores have been calculated they are aggregated to provide principle scores. For instance, the scores obtained by an animal unit for absence of injuries, absence of disease, and absence of pain due to management procedures are combined to reflect compliance of this unit with the welfare principle of 'good health'. The consultation with animal and social scientists had shown that they consider some criteria to be more important than others (e.g. in most animal types 'absence of disease' is regarded as more important than 'absence of injuries' which in turn is thought to be more important than 'absence of pain induced by management procedures'). Nevertheless the scientists did not want to allow compensation between the scores for these criteria (e.g. absence of disease does not compensate for the presence of injuries and *vice versa*). A specific mathematical operator (Choquet integral) was applied to enable us to take into account both of these lines of reasoning. Briefly, the Choquet integral is an aggregation function that generalises the notion of weighted average when weights are not only attached to each criterion but also possibly to any subset of criteria (Grabisch and Roubens, 2000). The Choquet integral calculates the differences between the minimum score and the next minimum score and assigns a weight (called 'capacity') to that difference. This process is repeated all the way through to the highest score. The capacities are calculated so as to minimise the sum of squares of the difference between the score calculated with the Choquet integral and the scores attributed by the persons consulted. An example of the calculation of principle-scores is provided in Box 7.4.

From welfare principles to overall assessment of farms

The scores obtained by an animal unit on all four of the welfare principles are used to assign that farm to a particular welfare category. How many and which welfare categories are necessary depend on the purposes for which the welfare assessment will be used. After consultation with the Advisory Committee of Welfare Quality®, various uses of the assessment were identified (Table 7.3) and four welfare categories were distinguished accordingly; these are:
- excellent: the welfare of the animals is of the highest level;
- enhanced: the welfare of animals is good;
- acceptable: the welfare of animals is above minimal requirements;
- not classified: the welfare of animals is low and considered unacceptable.

Box 7.4. Use of a Choquet integral to calculate the principle-scores for 'Good health'.

The 'Good health' principle integrates 3 criteria; 'absence of injuries', 'absence of disease', and 'absence of pain due to management procedures'. First, the scores obtained by a farm for the 3 criteria are sorted in increasing order. The first criterion-score is considered then the difference between that score and the next criterion-score is multiplied by the weight (called 'capacity' in a Choquet integral) of the group made of all criteria except the one that brings the lowest score. Following this, the difference between the last but one score and the next score is multiplied by the 'capacity' of the group made by the combined criteria except those that bring the two lowest scores. This can be written as follows:

$$\text{Principle-score} = \begin{cases} S_6 + (S_7 - S_6)\mu_{78} + (S_8 - S_7)\mu_8 & \text{if } S_6 \le S_7 \le S_8 \\ S_6 + (S_8 - S_6)\mu_{78} + (S_7 - S_8)\mu_7 & \text{if } S_6 \le S_8 \le S_7 \\ S_7 + (S_6 - S_7)\mu_{68} + (S_8 - S_6)\mu_8 & \text{if } S_7 \le S_6 \le S_8 \\ S_7 + (S_8 - S_7)\mu_{68} + (S_6 - S_8)\mu_6 & \text{if } S_7 \le S_8 \le S_6 \\ S_8 + (S_6 - S_8)\mu_{67} + (S_7 - S_6)\mu_7 & \text{if } S_8 \le S_6 \le S_7 \\ S_8 + (S_7 - S_8)\mu_{67} + (S_6 - S_7)\mu_6 & \text{if } S_8 \le S_7 \le S_6 \end{cases}$$

where S_6, S_7, and S_8 are the scores obtained by a given farm for Criterion 6 (Absence of injuries), 7 (Absence of disease), and 8 (Absence of pain due to procedures),
μ_6 μ_7 μ_8 are the capacities of Criterion 6, 7 and 8,
μ_{67} is the capacity of the group made of criteria 6 and 7,
etc.

A so called 'outranking' method is used to assign farms to these categories. Briefly, aspiration values are defined for each category. They represent the goal that the farm should try to achieve in order to be assigned to a given category. Membership rules then have to be set that describe how close to the aspiration values a farm should be in order to be assigned to the given category aimed for.

The question of whether the aspiration values should be absolute or relative was discussed in depth. Absolute aspiration values mean that they are set whatever the results obtained by farms. This approach might be risky because absolute aspiration values may be either too difficult or too easy to achieve and thereby fail to discriminate between farms. Relative values correspond to a certain percentage of farms from a given population in which the system is used; e.g. one could decide that farms scoring in the top 10% of their population are considered excellent, whatever the exact level of welfare they have achieved. However, such relative values seem difficult to handle for a number of reasons: (1) there is a risk that an excellent ranking may sometimes be achieved even if the level of welfare is low; (2) the classification of farms

Table 7.3. Potential uses of the welfare assessment.

Potential uses of the welfare assessment		Welfare categories needed	
		No.	Description
Scenario 1	Standard for cross-compliance and future definition of a minimum animal welfare standard	2	Below/above minimum legal requirements or equivalent (if no legislation)
	Compulsory labelling defining several levels of welfare	4	Poor/normal/enhanced/excellent
Scenario 2	Setting welfare targets for farm certification schemes – *voluntary labelling*	3	Excellent/enhanced/other (i.e. insufficient to enter a certification scheme)
Scenario 3	Feedback to producers for monitoring the results of welfare improvement strategies Self-assessment tool	Several	Very poor/excellent level of welfare (with intermediate categories)
Scenario 4	Assessing new animal farming systems/ breeds Furthering research on animal welfare	Several	Very poor/excellent level of welfare (with intermediate categories)

would depend on the population observed, in other words a farm may be classified as excellent in a given population but only as enhanced or even just acceptable in another population; (3) the adoption of relative aspiration values could prevent the monitoring and identification of progress on farms, e.g. a farm may remain in the category given in a previous visit even though the welfare of the animals has improved in the interim. It was finally decided to use absolute aspiration values and to check that these were realistic according to the results obtained from farms visited during the Welfare Quality® project. In addition, membership rules were adjusted according to the results obtained at numerous farms (see below).

Because all the welfare principles were considered to have similar importance for the animal's welfare status, it was decided that the same aspiration values should be adopted for all of the four principles:

- A minimum aspiration value of 20 was set for the 'acceptable' category because a score of 20 has a specific meaning on the 0-100 scale, i.e. if a farm scores below 20 on one criterion or principle it runs a high risk of failing to meet legal requirements and of being excluded from any welfare certification scheme.
- A score of 55 was set as the minimum threshold for the 'enhanced' category, i.e. just above a value of 50 which corresponds to 'not good and not bad'.

- A score of 80 was set as the minimum threshold for the 'excellent' category (symmetrical to the value of 20 set for 'acceptable' categorisation).

7.3 Testing the scoring model using Welfare Quality® datasets

The scoring model was tested using the first dataset to become available in the project; this incorporated data collected from 69 dairy farms (20 in Germany, 24 in Austria, and 25 in Italy).

7.3.1 From measures to welfare criteria

For the calculation of criterion-scores for dairy cows, two consultations were conducted: one before and one after the data had been collected on farms. When the initial evaluations by scientists (experts) were compared with the actual distribution of data across the dairy farms visited in Welfare Quality®, it appeared that the scientists had been rather severe. Regarding lameness for instance, 50% of farms having more than 12% severely lame cows would have been given scores of 30 or below. The distribution of lameness scores was presented to the same scientists (Figure 7.2) and they were asked if they were willing to change their initial evaluations. Most of them

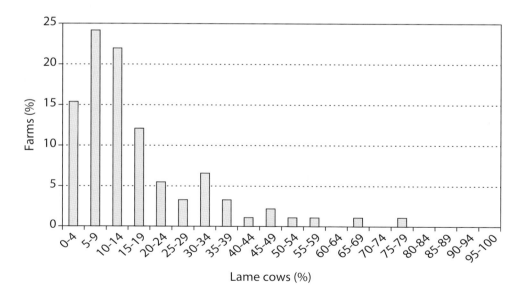

Figure 7.2. Distribution of lameness scores across dairy farms visited in Welfare Quality®. The percentage of lame cows is weighted for the severity of lameness. Weights: 0.29 for moderate and 1 for severe lameness.

were less severe during this second consultation (Table 7.4). However, since they considered lameness to be a serious problem for cows 50% of the farms still scored lower than 44. For the final scoring of farms, we used expert scores from the second consultation.

The calculations of scores for fattening cattle, veal calves, fattening pigs, and broilers were defined following just one consultation exercise because this was conducted after the data had been collected on farms.

7.3.2 From welfare criteria to welfare principles

For the calculation of principle-scores for dairy cows, two consultation exercises were again conducted: one before and one after data had been collected on farms. In this case, the two evaluations were almost the same. It was thus decided to proceed with only one consultation for the other animal types.

7.3.3 From welfare principles to overall assessment

The distribution of dairy farms across the aspiration values for the welfare categories 'acceptable', 'enhanced', and 'excellent' (respectively 20, 55 and 80) for each principle was then analysed. It turned out that it was extremely difficult for any farm to achieve a score of 80 on any welfare principle. Such a high value was obtained by a few farms for the 'Good feeding' and the 'Appropriate behaviour' principles. Nevertheless, a wide range of scores (from 1 to 100) was obtained. It was therefore decided that the three aspiration values should remain as defined, particularly bearing in mind that if the Welfare Quality® assessment system is put in place the post-evaluation implementation of welfare improvement strategies should be encouraged and thereby increase the likelihood that scores of 80 or more would be achieved.

Table 7.4. Comparison between scores assigned for lameness by animal scientists before and after being informed of the distribution of lameness across the dairy farms visited.

Lame cows (%)[1]	Mean expert scores 1st consultation	Mean expert scores 2nd consultation
10	33	48
30	14	21
60	6	6

[1] Weighted for severity; weight: 0.29 for moderate and 1 for severe lameness.

In order to guide decisions on whether or not a farm has performed well (or badly) enough to be considered to have fulfilled the conditions for a particular category, several membership rules were tested. A first, very intuitive rule is one of 'unanimity' which means that a farm needs to reach the aspiration value of a given category for all the welfare principles before it can be assigned to that category. For instance, to be categorised as enhanced a farm would need to score at least 55 on all four principles. Such a stringent rule did not seem realistic since half of the farms visited in the Welfare Quality® exercise described above would fall in the 'not classified' (unacceptable) category while the other half would be deemed only 'acceptable'. We therefore investigated other rules whereby a farm needs to achieve a score that is higher than the aspiration value of a particular category only on some of the principles (3 or 2 out of the 4) while not falling below the aspiration value of the next lower category for any principle. The likely distribution of farms using such rules was analysed and the classification obtained by farms according to the various rules tested was compared to the general impression of observers who visited the farms (expressed on a Visual Analogue Scale). The most appropriate rule appeared to be the following:

> A farm is considered 'excellent' if it scores more than 55 on all principles and more than 80 on two of them. It is considered 'enhanced' if it scores more than 20 on all principles and more than 55 on two of them. Farms with 'acceptable' levels of animal welfare score more than 10 on all principles and more than 20 on three of them. Farms that do not reach these minimum standards are not classified. In addition, an indifference threshold equal to 5 is applied: For instance, a score of 50 is not considered to be significantly different from one of 55.

7.3.4 Back to ethics

Although ethical issues were not addressed directly, the people consulted in the above exercise made their own ethical decisions and these are reflected in the scoring model.

First, it was clear that more emphasis was placed on the animals in very poor condition than on those in good condition. This line of thinking is reflected in the calculation of criterion-scores. For instance, if the same importance was given to animals in very bad and to those in very good condition then a score of 50 would be assigned to a farm with 50% severely lame cows and 50% cows that were not lame. In contrast, a score of 50 is assigned to a farm where 9% of the cows are severely lame and the others are not lame (see figure in Box 7.2). Nevertheless, it was clear that the overall welfare of a group of animals was also considered to be important. In the lameness example, a farm where 5% of the cows are severely lame and 50% are moderately lame is given a score lower than a farm where 10% of the cows are severely lame and none are moderately lame. Similar results are found whatever the animal types and the criteria.

Therefore, the scoring model was designed to reflect a balance between priority given to animals in the poorest conditions and the overall welfare of the whole herd or flock.

As mentioned earlier, compensation is very limited between welfare principles, but some might still be possible in certain circumstances. As a consequence, a farm that scores less than 20 on one principle might have some chance to finally be considered acceptable if it receives high scores for the other three principles (see above). The partners of Welfare Quality® are aware that this suggestion was not in line with most peoples' perception of overall welfare (where all principles need to reach a certain minimum level). In that case it was thought that a rule should state that a farm should not be considered acceptable if it falls below a score of 20 on one principle. However, at present such a rule appears premature. It may be advisable to wait until the Welfare Quality® assessment system has been implemented for a few years by which time clear improvements in welfare status should have become apparent at the assessed farms. The scoring model could then perhaps be refined in order to accommodate the more stringent perception.

Finally, the scoring model reflects two major considerations: (1) theoretical expectations of what is poor vs. good welfare, as reflected in the definition of aspiration values for the various welfare categories (20 for 'acceptable', 55 for 'enhanced', and 80 for 'excellent'); and (2) what can realistically be achieved in practice, as reflected in the rules chosen to assign farms to a particular welfare category (e.g. a farm needs to be at least excellent on two principles and only enhanced on the other two to be considered excellent). The nature of the final scoring model therefore represents a balance between theory and pragmatism.

Most assessment systems used in certification or control schemes are based on a number of specific points/requirements that the farm must comply with. To the best of our knowledge, these systems generally produce a simple pass/fail answer so it might be difficult for a farmer who has 'failed' to figure out how far he/she was from achieving a pass. By incorporating four clear welfare categories or grades the Welfare Quality® assessment system is considered more likely to encourage farmers to improve their status in regard to animal welfare.

The development of the scoring model is described in greater detail elsewhere (Botreau *et al.*, 2007, 2008, 2009; Veissier *et al.*, 2011).

7.4 Results obtained on European farms and the value of feedback

In total some 700 farms were assessed using the Welfare Quality® protocols in studies of pigs (United Kingdom, Spain, Sweden), cattle (Austria, Germany, Italy) and poultry (United Kingdom, Italy, the Netherlands). This section describes some of our findings from these visits as well as some comments that were made regarding the practical application of the Welfare Quality® protocols on these farms. After this exercise had been completed, we also explored the ways in which the information that was produced from the Welfare Quality® assessments could be used to promote and support management decisions and practices that were likely to improve welfare.

7.4.1 Consistency of results obtained on European farms

Because two or three visits were made to many of the farms assessed in the pilot study, it was possible to examine (at a very basic level) the variability of the results obtained between the two visits. Of course, the interpretation of these findings should be viewed with care as each visit will have been subject to variations in climate, season, age of stock, etc. However, although analysis of data from only few visits is unlikely to give an accurate representation of the real variability over a longer (many years) time frame it can still provide a little guidance. An example of data from several visits is given in Table 7.5 for beef bulls and Tables 7.6 and 7.7 for pigs (health measures).

An overall consistency between the results of the two visits was apparent with the common findings remaining 'common' and those conditions that occurred less frequently remaining at a low level of incidence. However, some results did differ from visit to visit. For the beef bull pilot study, depending on the measures, correlations between the assessments carried out during two visits ranged from a r_S of 0.03 to 0.97, while the percentage of farms that received the same welfare judgement in each of the two assessments ranged from a minimum of 38.8% to a maximum of 100%. In a recent study that was carried out after the Welfare Quality® project had finished, the protocol was applied on a small sample of dairy farms repeatedly in winter. The average scores obtained from a combination of the three visits were significantly correlated with the scores obtained at each individual visit, suggesting that the scores obtained by a farm for the various criteria are stable over time (Mounier, personal communication). Of course this suggestion likely depends on the farm situation remaining stable too, i.e. that no major changes in housing, feeding, management etc. have occurred during the study period.

Distinctions which must be made regarding the variability of a measure across assessments are: (1) variation due to real differences in the occurrence or intensity of the criterion being measured because of changes in management or feeding or due to an outbreak of a specific disease; and (2) the effects of improvements in welfare status (and in the measure in question) which have resulted from the adoption of management or disease control measures by the farmer as a result of the information that he or she received during the feedback of the findings of the first assessment.

Table 7.5 Consistency of the results obtained during 3 visits to beef farms at intervals of 6 months. The farms were in Austria, Germany, and Italy.

Measure	Initial to interim assessment					Interim to final assessment				
	r_S[1]	effect of assessment and country[2]	variance within farm < between farms[3]	farms with same WJ[4] [%]	n	r_S[1]	effect of assessment[2]	variance within farm < between farms[3]	farms with same WJ[4] [%]	n
% of lean animals	0.29*	/	/	91.5	59	0.68***	/	/	91.8	49
number of animals per drinker	0.95***	CO	yes	89.8	59	0.91***	CO	yes	87.8	49
% of groups with only one drinker	0.94***	ns	yes	100	59	0.94***	ns	yes	95.9	49
% of groups with dirty drinkers	/	/	/	90.9	44	/	/	/	94.1	34
duration of lying down	0.70***	CO	yes	100	55	0.45**	ASS, CO	no	100	42
% of standing ruminating to all ruminating	0.32*	CO	yes	49.2	59	0.44**	ASS, CO	no	40.8	49
% of dirty animals	0.74***	CO	no	74.6	59	0.47***	CO	no	73.5	49
% of groups with panting animal	/	/	/	88.1	59	/	/	/	100	49
space allowance per animal. bulls <350 kg	0.65***	ASS	yes	69.2	39	0.7***	ns	yes	84.4	32
space allowance per animal. bulls >350 kg	0.84***	ns	yes	76.3	59	0.72***	ns	yes	71.4	49
% of lame animals	0.36**	CO	yes	83.1	59	0.39**	CO	no	91.8	49
% of animals with nasal discharge	0.48***	ASS, CO	yes	44.1	59	0.15	ASS	no	44.9	49
% of animals with hampered respiration	0.03	/	/	66.1	59	0.19	/	/	75.5	49
number of coughs per animal and hour	0.55***	ns	yes	76.3	59	0.27	ASS	no	38.8	49
% of animals with ocular discharge	0.73***	ASS, CO	yes	86.4	59	0.69***	ASS, CO	no	57.1	49
% of animals with bloated rumen	/	/	/	91.5	59	/	/	/	98	49
% of animals with diarrhoea	0.43***	/	/	66.1	59	0.63***	/	/	71.4	49

Table 7.5 Continued.

Measure	Initial to interim assessment					Interim to final assessment				
	r_s[1]	effect of assessment and country[2]	variance within farm < between farms[3]	farms with same WJ[4] [%]	n	r_s[1]	effect of assessment[2]	variance within farm < between farms[3]	farms with same WJ[4] [%]	n
% of dehorned animals[5]	no data					0.97***	/	/	93.5	46
% of tail-docked animals[5]	no data					/	/	/	93.5	46
number of agonistic interactions per animal and hour	0.74***	ASS, CO	yes	64.4	59	0.48***	CO	no	55.1	49
proportion of agonistic to all social interactions	0.77***	CO	no	71.2	59	0.64***	CO	no	63.3	49
Avoidance distance	0.66***	ASS, CO	yes	50.8	59	0.77***	ASS, O	yes	59.2	49
% of pens with animals tongue rolling	0.74***	ASS, CO	yes	64.4	59	0.70***	CO	yes	65.3	49
Qualitative behaviour assessment	0.68***	ns	yes	74.6	59	0.53***	ns	yes	67.3	49

[1] Spearman's correlation coefficient r_s. ***: $P<0.001$; **: $P<0.01$; *: $P<0.05$.

[2] Significant effect of assessment (ASS) and country (CO); ns: not significant; /: test not conducted due to non-fulfilment of model assumptions (no variance).

[3] Yes: variance within farms smaller than between farms (consistency acceptable); no: variance within farms not smaller than between farms (consistency not acceptable); /: test not conducted due to non-fulfilment of model assumptions (no variance).

[4] WJ: Welfare Judgement.

[5] Comparison of either initial or interim assessment and final assessment depending on time point of conducting the management questionnaire.

Table 7.6. Differences in the incidence of health measures between two visits to pig farms, 12 months apart, in Spain.

	Visit 1					Visit 2					change (Visit 2 - Visit 1)	
	Mean	Median	SD	Min	Max	Mean	Median	SD	Min	Max	Δ	%Δ
% poor body condition	0.2	0.0	0.9	0.0	1.4	0.1	0.0	0.4	0.0	1.6	-0.1	-39.0
% bursitis 1	60.4	60.8	6.9	40.9	83.2	47.9	49.2	15.9	14.1	76.3	-12.5	-20.7
% bursitis 2	6.9	5.7	0.3	0.0	22.5	4.0	3.2	3.9	0.0	15.5	-3.0	-42.7
% total bursitis	67.3	67.0	7.1	40.9	94.7	51.9	51.9	18.3	14.1	79.9	-15.4	-22.9
% dirty 1	11.0	6.6	0.1	0.0	42.1	7.5	6.2	6.9	0.0	26.6	-3.5	-31.8
% dirty 2	3.6	0.8	18.0	0.0	29.0	2.3	0.0	5.7	0.0	19.7	-1.2	-34.0
% total dirty	14.5	8.0	17.9	0.0	54.1	9.8	7.0	10.7	0.0	39.8	-4.7	-32.3
% wounds 2	1.0	0.7	0.7	0.0	4.2	1.1	0.3	2.3	0.0	9.3	0.2	18.6
% tail 2	0.6	0.0	0.4	0.0	3.1	0.4	0.0	1.1	0.0	3.8	-0.2	-28.0
% lame 1	0.1	0.0	0.1	0.0	0.8	0.1	0.0	0.3	0.0	0.8	0.0	-6.4
% lame 2	0.2	0.0	0.1	0.0	1.7	0.1	0.0	0.3	0.0	1.3	-0.1	-57.2
% total lame	0.3	0.0	0.2	0.0	2.5	0.2	0.0	0.4	0.0	1.3	-0.1	-35.4
% breathing difficulty	0.1	0.0	0.1	0.0	0.7	0.0	0.0	0.0	0.0	0.0	-0.1	-100.0
% twisted snout	0.0	0.0	0.0	0.0	0.0	0.0	0.0	0.0	0.0	0.0	0.0	-
% prolapse	0.0	0.0	0.0	0.0	0.0	0.0	0.0	0.2	0.0	0.7	0.0	-
% total skin conditions	8.7	7.8	1.5	0.0	31.0	2.0	1.7	1.7	0.0	6.3	-6.7	-76.8
% hernia 2	0.1	0.0	0.1	0.0	0.9	0.1	0.0	0.3	0.0	0.9	0.0	32.8

Table 7.7. Spearman's rho correlation coefficients between some health measures taken during two visits of pig farms, 12 months apart, in Spain.

	% poor body condition (visit 1)	% bursitis – score 1 (visit 1)	% bursitis – score 2 (visit 1)	% total bursits (visit 1)	% dirty – score 1 (visit 1)	% dirty – score 2 (visit 1)	% dirty – total (visit 1)	% body wounds – score 2 (visit 1)
% poor body condition (visit 2)								
Correlation Coefficient	-0.194							
Sig. (2-tailed)	0.488							
% bursitis – score 1 (visit 2)								
Correlation Coefficient		0.304						
Sig. (2-tailed)		0.271						
% bursitis – score 2 (visit 2)								
Correlation Coefficient			-0.161					
Sig. (2-tailed)			0.567					
% total bursitis (visit 2)								
Correlation Coefficient				0.304				
Sig. (2-tailed)				0.271				
% dirty – score 1 (visit 2)								
Correlation Coefficient					0.590*			
Sig. (2-tailed)					0.021			
% dirty – score 2 (visit 2)								
Correlation Coefficient						0.019		
Sig. (2-tailed)						0.947		
% dirty – total (visit 2)								
Correlation Coefficient							0.522*	
Sig. (2-tailed)							0.046	
% body wounds – score 2 (visit 2)								
Correlation Coefficient								0.129
Sig. (2-tailed)								0.647

* Correlation is significant at the 0.05 level (2-tailed).

Given the relatively short time scale (18 months) in which these pilot studies were carried out in the Welfare Quality® project it is not yet readily apparent whether any significant changes seen between visits 1 and 2 reflect natural variability in the herd/flock performance or the impacts of management changes. The continued use of outcome based measures will provide larger longitudinal data sets which will in turn provide increasing clarity on which measures provide reliable data with long term collection, and which ones are the most likely to drive change and improvement in welfare status through the monitoring of specific outcome measures and through the use of aggregated farm scores.

7.4.2 The value of feedback of Welfare Quality® results to the farmers

The assessment of welfare and the feedback of results are vital first steps in any effort aimed at helping producers improve the well-being of their animals. The circle should then be completed by the provision of advice on remedial solutions to problems and/or risk factors identified during the assessment (see Chapters 8 and 9).

For present purposes the assessment of pododermatitis is taken as an illustrative example.

Foot pad dermatitis (or pododermatitis) in broiler chickens is a contact dermatitis found on the skin of the foot, most commonly on the central pad, but sometimes also on the toes. The skin is turned dark by contact with wet litter (often with high pH) and deep lesions can result. Foot pad lesions are common, but their prevalence is variable with some flocks showing much higher levels than others. The Welfare Quality® assessment allows recording not only of the proportion of the flock affected but also of the severity of the lesions.

As a first step, the producer (or an operative of the assurance or advisory service, or the veterinarian normally used by the producer) could determine the prevalence and severity of foot pad dermatitis within similar flocks (e.g. linked to the same company). For instance, inspection bodies in some countries are now beginning to focus on foot pad health as a marker for company welfare performance. It should also be possible to estimate the economic impact of lameness in terms of the numbers of small, moribund and culled birds. In general, a company can realise significant improvements in profitability and overall bird welfare if foot pad problems are tackled appropriately.

Second, within a given company, comparisons between 'good' and 'poor' farms (with respect to foot pad dermatitis) can be made to help identify management, house environment, feeding, medication, stockmanship and genotype factors which differ between these farms and which may be linked to the foot pad problem in the poor farms. In parallel, the usage of water should be checked, since farms with increased water use per bird (in equivalent weather conditions) may have systematic problems with leaking drinkers. This is important because chronic leakage of even small amounts of water into the litter can severely damage its management and, consequently, foot pad health. In addition, an investigation of the bacteriological pathologies linked with foot pad dermatitis can help determine whether these bacteria originate in the hatchery, in the transportation stage, or through lapses in farm biosecurity.

Third, the farmer could be informed about the extent of foot pad dermatitis on his/her farm, and, with time and appropriate analysis, a pattern of risk factors may emerge which can guide his/her decisions on how to reduce the problem. In real farm experience risk factors for pododermatitis and lameness are known to include: growth rate, the age to which the birds are grown and slaughtered, the use of whole cereals in the diet, the type of feed, the quality of biosecurity measures, litter condition (a very important factor), and the genotype of the birds. Additionally, the gender of the birds, the levels of feed restriction, lighting pattern, light intensity, bird activity levels and stocking density have been manipulated on commercial farms in efforts to control the levels of foot pad dermatitis in chickens (Butterworth, 2009; Faure *et al.*, 2003).

Another example of the value of providing helpful feedback to producers emerged from the pilot studies carried out in pigs and is shown in Figure 7.3.

7.4.3 Responses to feedback of the Welfare Quality® results

After the first assessment visit, some of the farmers were given advice (as described above) whereas others were not. The subsequent comparison of the two groups enabled us to establish whether or not such feedback can encourage the farmer to make the necessary changes to secure improvements in welfare. Some examples are given below.

Regarding foot pad dermatitis in broilers, although there was some scope for improving the welfare status of the animals on all the farms visited we found no significant effect of providing feedback and advice. In fact, only a minority of farmers that had received feedback had actually implemented any of the advised measures. The reasons given for this lack of action varied, but included financial and practical aspects as well as a lack of motivation because the farmers did not perceive the situation as needing improvement. Of course, the strength of this comparison may have been weakened by the relatively short time frame of the Welfare Quality® project and a consequently limited opportunity for the farmer to implement suggested improvements. An even more influential demotivator may have been the lack of incentives in terms of

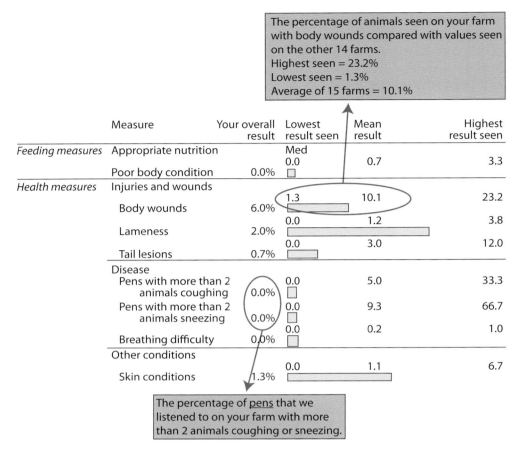

Figure 7.3. Example of a pig farm 'feedback' report.

certification and of any perceived possibility to achieve higher prices. Encouragingly though, those farmers that had implemented some of the suggested changes mostly perceived them as having reduced the pododermatitis problem and thereby improved the welfare of their stock.

Eight out of 20 farmers involved in the beef cattle studies reported that they had applied or adopted some changes following the post-assessment provision of advice. More specifically, the following suggestions were implemented:
- modifying feeding to reduce diarrhoea;
- installing water bowls as well as nipple drinkers (though this action followed the final rather than the first assessment);
- use of local anaesthetic by farmers when dehorning calves;

- reducing the stocking density (done by three farmers; to provide better access to drinkers in one case);
- improving the air flow in the barn to reduce ocular discharge;
- adding a portable air fan to improve ventilation;
- improving the cleanliness of the barn;
- adding one drinker for bulls <200 kg (although not in the primary scope of the intervention).

The following reasons were given as to why advice was not implemented by the farmers:
- high costs (two farmers);
- no time (two farmers);
- installing additional drinker(s) is difficult (two farmers);
- no motivation to change anything because the farmer is content with the barn (five farmers);
- reduction of animals per pen is not perceived as profitable and would increase problems with the sloped floor (too much litter);
- improvement strategies seemed to be potentially costly or to be difficult to implement in the short term (mentioned by two farmers);
- spending more time with animals would not improve the animal-human-relationship in older animals;
- changes would be made if or when the barn was to be rebuilt;
- one farmer did not consider cessation of dehorning to be an acceptable alternative to current practice, and dehorning is already done with anaesthetics.

7.5 Conclusion

A particularly important objective of the researchers was to ensure that the Welfare Quality® assessment protocols should provide a way of reliably assigning farms to one of the four welfare categories described above. To support this, a software package was designed to ease collection of data from the 30-50 measures taken per animal type (using a laptop or tablet PC), for subsequent storage in a database, and the calculation of welfare scores at criterion, principle and overall levels. An interactive platform which enables dialogue with stakeholders has also been established (www.clermont.inra.fr/wq/).

The four welfare categories (not classified, acceptable, enhanced and excellent) can be used for a number of different purposes. Some, such as assisting or ensuring compliance with legislation or with the requirements of a certification scheme, only really need two categories. Thus, in the legislative context one must simply determine if the farm is at least acceptable versus not classified. Similarly a certification scheme may require farms to secure placement on the enhanced or excellent categories and could reject farms which are classified below. In contrast, for the provision of feedback to producers or for the assessment of new farming systems and breeds it would be much more meaningful and accurate if the classification of farms or systems made use of the full range of categories as well as more detailed underlying data (see Chapter 9).

The integration process was built upon the results of and experience gained during an iterative exercise of scoring virtual farms. This was done both to estimate welfare scores at criterion level from the results obtained at the measures level, and then to synthesise criterion scores into principle scores. In order to then synthesise principle scores into an overall welfare assessment, aspirational values (minimum thresholds for various categories of welfare status) were chosen after extensive discussion between scientists and stakeholders. Recent research showed that society at large may not view the four welfare principles defined in Welfare Quality® as equally important, e.g. behaviour is viewed as less important than health (Tuyttens *et al.*, 2010). This perception was not only recognised in the early stages of the project but it was also incorporated in the scoring process (see above). For instance, although the four Welfare Quality® principles are given the same importance in the final classification of farms, the actual construction of criteria varies between principles so that the scoring is more severe for health issues than for behavioural problems. As a consequence, health problems impact more on the final classification of farms than do behavioural problems (Botreau and Veissier, 2011). A lesson learnt in this context is that Welfare Quality® may not have succeeded in communicating this information clearly enough. This message has been taken on board (see concluding paragraph).

Another issue with regard to the integration process was that it had not been possible to identify farms that scored highly enough to be placed in the 'excellent' category among the population of animal units surveyed during the Welfare Quality® project. We therefore wondered if the aggregation process was too severe so we re-examined the best farms. This exercise confirmed the original finding that none of them could be qualified as excellent. However, since the main purpose of the survey had been to establish that the measures incorporated in the protocol did enable the detection of welfare problems, the farms and the timing of the visits had been chosen so as to maximise the likelihood of observing measurable problems; for instance dairy farms were visited at the end of winter. Encouragingly, later studies using the Welfare Quality® protocols have revealed that although no excellent farms were identified the proportion of 'enhanced' farms was higher than that found in the original population surveyed during the project (De Vries *et al.*, 2011). Once the protocols have been used on larger populations of farms, we will be able to propose appropriate refinement of the aspiration levels or of the rules for allocating dairy or other farms to the various welfare categories.

The scoring system remains rather complex because it aims to reflect how people reason about animal welfare. It is thus important that the users of the system are able to clearly understand how the system works and what kind of results it can produce. Continuous clear communication about the scoring system is necessary to assure its full understanding by potential users (Veissier *et al.*, 2011). This could take the form of individual contact, workshops and/or fact sheets. In this context, Welfare Quality® fact sheets were very well received by a wide range of stakeholders. Encouragingly too, when applied on farms the Welfare Quality® protocols were rather well viewed by the farmers who perceived them as trustworthy and interesting.

References

Bekoff, M. (2008). Why 'good welfare' isn't 'good enough': minding animals and increasing our compassionate footprint. ARBS Annual Review of Biomedical Sciences, Theme Topic on 'Unraveling Animal Welfare', 10, T1-T14.

Botreau, R., Bracke, M.B.M., Perny, P., Butterworth, A., Capdeville, J., Van Reenen, C.G. and Veissier, I. (2007). Aggregation of measures to produce an overall assessment of animal welfare: Part 2 – Analysis of constraints. Animal, 1, 1188-1197.

Botreau, R., Capdeville, J., Perny, P. and Veissier, I. (2008). Multicriteria evaluation of animal welfare at farm level; an application to MCDA methodologies. Foundations of Computing and Decision Sciences, 33, 287-316.

Botreau, R. and Veissier, I. (2011). Priorities between welfare issues: experts' choices made in the Welfare Quality®project. In: UFAW (ed.) UFAW International Symposium 2011: Making animal welfare improvements: economic and other incentives and constraints, UFAW, London, UK.

Botreau, R., Veissier, I. and Perny, P. (2009). Overall assessment of cow welfare: strategy adopted in Welfare Quality®. Animal Welfare, 18, 363-370.

Bracke, M.B.M., Spruijt, B.M. and Metz, J.H.M. (1999). Overall animal welfare assessment reviewed. Part 1: Is it possible? Netherlands Journal of Agricultural Science, 47, 273-291.

Butterworth, A. (2009). Animal welfare indicators and their use in society. In: Smulders, H. and Algers, B. (eds.) Welfare of production animals: assessment and management of risks. Food Safety Assurance and Veterinary Public Health Series no. 5. Wageningen Academic Publishers, Wageningen, the Netherlands, pp. 371-389.

De Vries, M., Van Schaik, G., Bokkers, E.A.M., Dijkstra, T. and De Boer, I.J.M. (2011). Characteristics of dairy herds in different Welfare Quality catergories. In: Widowski, T., Lawlis, P. and Sheppard, K. (eds.), Proceedings of 5th international conference on the assessment of animal welfare at farm and group level. Wageningen Academic Publishers, Wageningen, the Netherlands, p. 9.

Faure, J.M., Bessei, W. and Jones, R.B. (2003). Direct selection for improvement of animal well-being. In: Muir, Aggrey, E. (eds.) Poultry breeding and biotechnology. CAB International, Wallingford, UK, pp. 221-245.

Fraser, D. (1995). Science, values and animal welfare: exploring the 'inextricable connection'. Animal Welfare, 4, 103-117.

Grabisch, M. and Roubens, M. (2000). Application of the Choquet integral in multicriteria decision making. In: Grabish, M., Murofushi, T. and Sugeno, M. (eds.) Fuzzy measures and Integrals – theory and applications; studies in fuzziness and soft computing. Physica-Verlag, Heidelberg, Germany, pp. 348-374.

Holm, L., Ritz, C. and Ladewig, J. (2007). Measuring animal preferences: shape of double demand curves and the effect of procedure used for varying workloads on their cross-point. Applied Animal Behaviour Science, 107, 133-146.

Miele, M., Veissier, I. and Evans, A. (2011). Animal welfare: establishing a dialogue between science and society. Animal Welfare, 20, 103-117.

Tuyttens, F.A.M., Vanhonacker, F., Van Poucke, E. and Verbeke, W. (2010). Quantitative verification of the correspondence between the Welfare Quality® operational definition of farm animal welfare and the opinion of Flemish farmers, citizens and vegetarians. Livestock Science, 131, 108-114.

Veissier, I., Jensen, K.K., Botreau, R. and Sandøe, P. (2011). Highlighting ethical choices underlying the scoring of animal welfare in the Welfare Quality® scheme. Animal Welfare, special issue *Knowing animals,* 20, 89-101.

Whay, H.R., Main, D.C.J., Green, L.E. and Webster, A.J.F. (2003). An animal-based welfare assessment of group-housed calves on UK dairy farms. Animal Welfare, 12, 611-617.

Chapter 8. Welfare improvement strategies

Xavier Manteca and Bryan Jones

8.1 Introduction

The welfare assessment protocols that were developed for cattle, pigs and poultry in the Welfare Quality® project (and described in Chapters 5 and 6) clearly provide extremely important tools that can help the farmer to improve welfare management on the farm. At an early stage during the project´s inception it was also recognised that to support such management systems a concerted European effort in the area of animal welfare should include research designed to identify practical ways of solving or at least alleviating some of the main welfare problems in current animal production.

Animal welfare is a complex and multifactorial concept; an animal´s welfare status is determined by both internal and external variables, and the causes of welfare problems can be numerous and varied. Firstly, for example, many farm animals, particularly poultry and pigs, may be kept under relatively low levels of sensory input in some modern farming systems; this is likely to engender negative emotional states such as fear, frustration, apathy and the development of injurious behaviours (Jones, 1997, 2001; Mench, 1994). The elicitation of such negative states can also result in undesirable economic consequences, for example they can reduce productivity, product quality and profitability (Jones and Boissy, 2011). Secondly, environmental constraints may prevent the animal from adopting a suitable response to a challenging event, for example a caged hen or tethered cow is unable to run away from a threatening stimulus (Jones, 1998). The inability to respond appropriately to challenge can further engender fear, distress, frustration, depression and other undesirable states. Thirdly, it is widely accepted that farming practice and selective breeding have often changed too rapidly and frequently for the animals' biology and behaviour to evolve appropriately and at the same pace (Faure *et al.*, 2003; Jones and Hocking, 1999). Potential solutions may lie in identifying suitable changes to the environment and/or the animal (Faure and Jones, 2004; Faure *et al.*, 2003).

During the past two decades it has become increasingly recognised that appropriate environmental enrichment, including positive human contact, can dramatically enhance animal welfare (Hemsworth and Coleman, 1998; Jones, 1998, 2001, 2004; Mench, 1994) and that selective breeding is also a powerful tool for alleviating welfare problems (Faure *et al.*, 2003; Grandin and Deesing, 1998; Jones and Hocking, 1999). Of course, any proposed 'welfare-friendly' changes to housing and husbandry systems or to breeding programmes must be shown to be practicable, safe, desirable and affordable. Within these constraints a major sub-project within the Welfare Quality® project was designed to develop and validate innovative, knowledge-based and practical species-specific strategies for improving on-farm animal welfare.

Collectively and in general terms, the researchers in this sub-project aimed at minimising the elicitation and expression of damaging behavioural and physiological traits and states, improving the human-animal relationship, and providing the animals with safe and stimulating environments. Initially, they considered several key welfare problems that are perceived as important by a broad range of European stakeholders, including producers, processors, retailers, academics, government, non-governmental organisations and members of the public. The main criteria used to identify the 'target problems' were as follows:
- Their alleviation could greatly improve the quality of life for farm animals as well as often improving productivity and product quality and thereby generating economic benefits for the farmers.
- The problem has a substantial impact on any one or more of the Four Principles of Good Welfare identified in the Welfare Quality® Project.
- High-quality scientific research was urgently needed in that particular area

Based on the above criteria a joint exercise involving biologists and social scientists resulted in the formulation of six work packages (WP), each of them addressing one of the following welfare issues:
- WP 3.1: Minimising handling stress in pigs, poultry and cattle: improving stockmanship.
- WP 3.2: Identifying genetic solutions to specific welfare problems: leg conformation and longevity in pigs, and psychobiological characteristics and adaptation in dairy cattle.
- WP 3.3: Eliminating injurious behaviours (feather pecking and cannibalism in laying hens, and tail biting in pigs).
- WP 3.4: Reducing lameness in broiler chickens and dairy cattle.
- WP 3.5: Minimising neonatal mortality in pigs.
- WP 3.6: Alleviating social stress in pigs and intensively kept beef cattle.

The approaches adopted in the various work packages included the development of new management/husbandry practices (WPs 3.1, 3.6), environmental manipulation (WPs 3.4, 3.6) and potential genetic strategies (WPs 3.2, 3.5, 3.6). In the present chapter we describe the rationales, approaches, results and implications of five research projects contained within the overall programme of the work packages. These were specifically chosen to represent each of the above approaches (management, environmental and genetic) and each of the main species (cattle, pigs and poultry).

They include: stockmanship improvements, lameness in broiler chickens, neonatal mortality in piglets, social stress in beef cattle and aggression in pigs.

8.2 Reducing handling stress and improving stockmanship

The term 'stockmanship' covers the way(s) in which animals are handled, the quality of their daily management and their health care (Waiblinger and Spoolder, 2007). At least three factors underlie individual differences in the quality of stockmanship, these are the stockperson's personality, attitude and behaviour (Hemsworth and Coleman, 1998; Jones, 1996). Personality is a relatively stable characteristic over time and it has been defined as the unique combination of traits that affects how a person interacts with the environment. Attitudes (including those towards animals) are learnt and may be modified through experience and education; they are often regarded as the most important factor explaining how a person interacts with social objects, including animals (see Waiblinger and Spoolder, 2007 for summary and references). More specifically, both personality and attitudinal factors influence the way that stockpersons behave towards the animals in their care.

The quality of stockmanship has a profound effect on the animals' welfare and productivity (Boivin *et al.*, 2003; Hemsworth and Coleman, 1998). For instance, despite centuries of domestication exposure to human beings is still one of the most potentially alarming events encountered by many farm animals. Indeed, unless they are accustomed to human contact of either a neutral or positive nature the predominant reaction to people is one of fear (Duncan, 1990; Jones, 1997). Not surprisingly, the problem is exacerbated by exposure to rough, aversive and/or unpredictable handling. Indeed, many human-animal interactions in current farm practice are inherently frightening, e.g. restraint, depopulation, beak trimming, disbudding, forced movement, loading and unloading, veterinary treatment, while few, other than feeding, are positively reinforcing. We must also remember that contact with humans could become even more stressful if increasing automation results in reduced opportunities for animals to habituate to people. The stockpersons' behaviour, which can vary from calm, gentle, frequent and 'friendly' to infrequent, rough and rushed, is a major variable determining animals' fear of or confidence in humans and, hence, the quality of the human-animal relationship (Hemsworth and Boivin, 2011). Chronic fear of humans is a major problem that can cause handling difficulties, injury (to the animals and/or the stockperson) and stress as well as impaired growth, poor reproductive performance and reduced product quality in a range of farmed animals including laying hens, broiler chickens, pigs and dairy cattle (Hemsworth and Boivin, 2011; Hemsworth and Coleman, 1998; Jones, 1997). For example, a series of studies found negative (and probably causal) correlations between fear of humans and productivity in the dairy, egg, broiler and pig industry

(Hemsworth, 2003). Conversely, experience of positive human-animal interactions has been shown to decrease the animals' general level of fear and distress (Jones 1993, 1995; Seabrook and Bartle, 1992) and enhance reproductive performance (Waiblinger *et al.*, 2006).

A strong influence of the stockpersons' attitudes on various aspects of farm animal welfare and production has been reported in several species (Boivin *et al.*, 2003; Hemsworth, 2003). For instance: (1) attention to detail is essential in a farrowing house to reduce neonatal mortality in piglets (Holyoake *et al.*, 1995); (2) farmers' attitudes to dairy cows and the degree to which their housing was designed and managed in order to fulfil the animals' needs are positively correlated (Mülleder and Waiblinger, 2004 in Waibingler and Spoolder, 2007); (3) a good attitude is associated with increased contact which, in turn, improves the stockperson's knowledge of the animals and facilitates the early recognition and solution of problems (Waiblinger *et al.*, 2006).

Cognitive-behavioural intervention techniques designed to target and improve those attitudes and behaviours of stockpersons that have a direct effect on the animals' fear of human beings and their general welfare (Hemsworth, 2003) show undoubted promise. Indeed, successful educational/training programmes have already been developed in Australia (Hemsworth and Coleman, 1998). Their relevance to the European context was subsequently evaluated in the Welfare Quality® project with a view to developing a suitable multi-media training programme for stockpersons.

8.2.1 Welfare Quality® studies

Welfare Quality® researchers carried out several studies designed to increase general understanding of stockmanship. For example, a questionnaire focusing on animal handling problems was mailed to 300 beef cattle farmers/breeders in France; this covered the perceived ease of handling the cattle, husbandry conditions, farmers' attitudes towards cattle and their behaviour during husbandry and handling procedures. Some breeders were then visited and interviewed, and their calves' behaviour was observed in a crush test (where animals are restrained individually in a specific apparatus) in the presence of a human. The researchers also observed numerous transfers (including loading and unloading) of beef bulls from commercial farms to a slaughter plant, measured plasma cortisol concentration and meat pH in the carcasses, and they studied the farmers' attitudes towards bulls and to working with them. Dairy cattle husbandry was also examined in randomly chosen farms in Austria and Italy. This effort involved surveys of farmer attitudes and behaviour as well as direct assessment of the cows' reactions to humans in standardised tests (with or without prior handling) and of the stockpersons' behaviour when milking.

Both beef and dairy farmers emphasised the importance of good human contact (quality and frequency) and of good facilities in increasing the ease of handling. 28% of farmers were not aware that genetic background is important in determining the ease of handling. This is particularly surprising since the temperament of heifers or cows was the first trait they considered in decisions on culling. Farmers showed some negative behaviours (hitting, shouting) in certain situations but their attitudes towards such behaviours were independent of those towards animals.

Calves were much calmer if the farmers enjoyed contact with their animals than if they had little interest in them. Interestingly though, the beneficial effects of a gentle handling regime were only retained when the dam herself was docile. Dairy cows that had received positive human contact approached closer to an unfamiliar human at test and most dairy farmers agreed that calm, gentle and patient handling is important. However, nearly 20% of them felt that cows should be fearful of humans in order to make them easier to handle. Calves which were reared outdoors, separated from their dam each day and gently handled during the first weeks of age were consistently and durably (up to 40 weeks) less fearful of humans than non-handled ones. Beef bulls from farms where the farmers had positive attitudes towards them showed lower levels of the stress hormone cortisol and better meat quality after transport. Paradoxically though, handling and loading beef bulls prior to transport was more difficult if they had received regular human contact. Loading beef bulls into the truck was easier on farms equipped with a corridor or a loading ramp, thus underlining the importance of using appropriate equipment. Unloading was easier when the journey was short and when the local temperature was high.

In another experiment dry and lactating Holstein Friesian cows were positively handled for brief periods on each of 3 days per week for 4 weeks, whereas similar numbers received no handling and could neither see nor interact with the treated animals. In subsequent tests the handled cows came closer to both familiar and unfamiliar persons (i.e. reduced fear of humans), and the handling treatment was more effective if it was applied during the dry period than during lactation.

Finally, Welfare Quality® researchers developed a multi-media stockmanship training programme. This task combined and built on existing literature, the data and material (photographs, videos, technical reports, etc.) generated in related studies, and the team's collective experience of training to develop an effective multi-media training programme for stockpersons working with cattle, pigs and poultry. Inter-continental collaboration also married key expertise from Europe and Australia. First, the team defined its objectives (prepare a script, story board, visual aids, etc.), methodologies (e.g. literature review, questionnaires) and agreed on a common structure (even if some aspects varied across species). It was also agreed that the training package should

utilise cognitive-behavioural intervention techniques to specifically target those key attitudes and behaviours of stockpersons that are known to have a direct effect on farm animals' fear of humans. The 'Quality Handling' programme (software, trainer manuals, newsletters, etc.) has been developed and tested in the various species. This programme, which is now commercially available, covers the following issues:
- How animals' fear responses to humans vary between farms.
- How fear of humans can adversely affect productivity and ease of handling.
- How animals perceive their environment.
- How to build a positive human-animal relationship.
- How to improve the stockpersons' attitudes and behaviour towards the animals.
- How to maintain the above improvement when the stockpersons return to the farm after training.

8.3 Reducing lameness in broiler chickens

Lameness resulting from leg disorders is commonly regarded as one of the main welfare problems in broiler chickens (FAWC, 1992, 1998; European Commission, 2000). Leg problems have serious consequences for welfare because lame birds may suffer chronic pain (Pickup *et al.*, 1997) and their behavioural repertoire can be significantly restricted, e.g. they may have difficulty accessing feeders and drinkers (Weeks *et al.*, 2000). Lameness can also have significant economic costs, some birds may have to be culled and the surviving lame birds may lose weight and are more likely to be downgraded at slaughter (Kestin *et al.*, 1999).

As many as 90% of birds in some flocks are thought to show at least some degree of lameness by the time they reach slaughter age (Kestin *et al.*, 1992), and some studies report that up to 30% of the birds were moderately to severely lame (Sanotra *et al.*, 2001). However, the prevalence of lameness in broilers varies considerably between farms. A large-scale study (Dawkins *et al.*, 2004) found a mean percentage of severely lame birds of 9%, with a range of 0 to 20. As intensive broiler chicken production now exceeds 2×10^{10} birds worldwide (Dawkins *et al.*, 2004), lameness in broilers is likely to be one of the most widespread farm animal welfare problems in modern agriculture. Despite this information farmers are often thought to significantly underestimate the amount of lameness in their broiler flocks and in doing so they risk compromising the birds´ welfare as well as product quality and profitability.

The aetiology of leg disorders in broilers includes many factors such as genetic background, gender, growth rate, feed conversion efficiency, body conformation, exercise, nutrition, stocking density, climate and management. These categories are not mutually exclusive because one factor may affect another (Bradshaw *et al.*, 2002). Leg disorders can be classified according to their underlying pathology as infectious, developmental and degenerative (Bradshaw *et al.*, 2002), with tibial dyschondroplasia and long bone deformities being particularly common (Julian, 1998). Over the last 40 years, genetic selection for rapid growth and improved feed conversion efficiency, together with changes in the feed encouraging high nutrient intake, have markedly increased growth rate which, in turn, has been implicated in the increasing prevalence of leg problems (Bradshaw *et al.*, 2002; Julian, 2005; Kestin *et al.*, 2001; Sanotra *et al.*, 2003). In other words, the birds' growth rate exceeds that of their skeletal development. Higher growth rates may also predispose birds to bacterial infection (Corr *et al.*, 2003; McNamee *et al.*, 1999). Leg weakness is positively correlated with live weight gain and is more pronounced in males than females (Kestin *et al.*, 1994; Sanotra, 2000; Sorenson *et al.*, 2000), possibly because of the sex differences in conformation and growth. It has also been suggested that selection for greater feed conversion efficiency has reduced the birds' performance of energy consuming behaviours; indeed locomotor activity is much lower during the finishing period in chickens from fast-growing genetic types than in slow-growing ones. In addition, the correlation between activity levels at early and later ages indicates the involvement of genetic factors in the expression of locomotor behaviour in very young chicks (Bizeray *et al.*, 2000). Metabolic imbalances induced by high nutrient intake may also cause some of the conditions that result in lameness and these might be corrected without reducing growth rate (Julian, 1998).

Although high stocking density was generally thought to be one of the major risk factors for lameness in broilers (Bradshaw *et al.*, 2002), it was recently reported that stocking density was, within limits, less important than other factors such as temperature, humidity and stockmanship (Dawkins *et al.*, 2004). Nevertheless, stocking density still had some effect, and at the highest densities more birds showed signs of lameness. Furthermore, activity is thought to be inversely related to stocking density, and increased activity reduces valgus/varus deformity.

8.3.1 Welfare Quality® research

Welfare Quality® researchers investigated whether specific changes to the diet and feeding regime could help to alleviate lameness in broiler chickens. They then used the results to develop an innovative feeding strategy. Unlike the traditional continuous provision of a starter diet followed by a finishing diet, this new strategy involved the sequential feeding of two diets varying in protein and energy content for a strictly delineated period in the broilers' lives. For the first seven days of life broiler chicks were fed a standard starter diet. Then from day 8 to day 28 the diets were rotated every 48 hours between: (1) a low energy, high protein (E-P+) diet consisting of 97% of the energy and 121% of the protein of a standard diet; and (2) a high energy, low protein (E+P-) one consisting of 103% of the energy and 79% of the protein content of a standard diet. Thus there were 10 cycles of E-P+ and E+P- diets. The birds were then given a standard finishing diet from day 29 till slaughter at day 38. A control group received the traditional feeding regime (see above).The broilers' ability to walk (gait score, GS) and their general performance (e.g. feed conversion, final weight) were then evaluated at the end of the experiment.

Gait score was found to be better in birds on the sequential diet regime than in the controls (mean GS of 2.41 and 2.61, respectively) and there were no significant treatment effects on body weight at slaughter. Neither feed conversion nor carcass conformation was impaired by sequential feeding, and an increase in abdominal fat was small enough to be avoided by improving diet composition. These findings strongly suggest that lameness in broilers can be alleviated by slowing down their early growth rate and speeding it up again once their bones have developed.

In short, this novel feeding regime not only reduced the instances of lameness but it also brought the broilers up to standard slaughter weight without the need for any additional feeding days. The researchers are still analysing the exact price differences between the broiler standard diet and the sequential diets, but initial results suggest that the latter were not more expensive than the standard diet. This sequential feeding method could be a win-win situation for the chickens and the farmers. In other words, it could improve the birds' welfare by reducing lameness at no extra cost while safeguarding the farmers' profits at the same time.

8.4 Reducing neonatal mortality in pigs

Pigs show a high prevalence of neonatal mortality. Data from the United Kingdom, for example, indicates that 11.85% of all live-born pigs die within the 72h post-parturition period (Meat and Livestock Commission, 2006). Besides constituting an important economic problem, piglet mortality is also becoming an increasingly significant welfare concern.

Neonatal mortality in pigs is a complex multi-factorial problem that involves several elements related to piglet health status and behaviour, the behaviour of the sow and the characteristics of the physical environment (Baxter *et al.*, 2008). Crushing is the most common and ultimate event preceding live-born death, although hypothermia and starvation are often underlying and important factors resulting in the piglet being more susceptible to challenge (Edwards, 2002).

The piglets' level of development and physical condition at birth has a major impact on survival. Live-born mortality is also highly dependent on the piglets' vigour, irrespective of its relation to body weight. Less active individuals face a higher risk of being crushed through a variety of interplaying factors. For example, it takes longer for them to locate the udder and to suck the colostrum, which in turn compromises weight gain and increases the risk of hypothermia and starvation. Hypothermic piglets tend to seek closer contact with the sow, thus raising their likelihood of being crushed. Indeed, crushing is more prevalent in outdoor (colder conditions) than in indoor herds (Edwards, 2002). Both the lack of vigour and a poor physical condition in newborn piglets are correlated with some physiological traits, such as rectal temperature, some laboratory measures, e.g. reduced plasma concentrations of urea, phosphor, calcium, and a poorer index of in vitro cellular immune function (Tuchscherer *et al.*, 2000).

Practical measures intended to reduce neonatal mortality have been mainly centred around modifications of the farrowing environment based on the different causes of piglet death. However, farrowing crates are known to stress the sow and may also be involved in the causes of other types of piglet mortality, such as savaging. Logically, the development and implementation of strategies designed to reduce hypothermia and starvation should decrease the incidence of piglet mortality. When the piglet is born and makes the transition from the thermoneutral intrauterine environment to the extra-uterine environment, it is exposed to a 15-20 °C drop in temperature (Herpin *et al.*, 2002), and, not surprisingly, providing additional heat sources at the birth site during farrowing can decrease mortality. For instance, Morrison *et al.* (1983) improved survival by providing heat lamps at the site of farrowing, a method that can be applied when the sow is restrained in a crate. However, farrowing sows in loose-housed accommodation require different methods of providing thermal comfort. For example, the provision of under-floor heating at the time of farrowing improved piglet survival; (Malmkvist *et al.*, 2006). Providing deep-straw bedding (a common practice in outdoor systems) can also help by reducing the rate of heat loss and thereby creating a more suitable microclimate (Wathes and Whittemore, 2006). Additional management strategies designed to decrease mortality include increased supervision and intervention at the time of farrowing to assist the birth process, and thereby limit the incidence of stillbirths, and to help weak piglets find the teat and suckle colostrum (White *et al.*, 1996). Additionally, many aspects of piglet survival are heritable and there is thought to be sufficient genetic variance in the population to allow economically viable selection for welfare-friendly characteristics (Knol *et al.*, 2002).

8.4.1 Welfare Quality® research

The work carried out by Welfare Quality® researchers had two main objectives:
- To identify behavioural and physiological characteristics of piglet survival.
- To consider the effects of genetic selection for survival in alternative farrowing systems to the conventional farrowing crate.

Stillborn mortality was found to be correlated with a reduced piglet body weight and, more precisely, with having a disproportionately long and thin body shape, abnormal shape proportions, and being born late in the farrowing birth order (Baxter *et al.*, 2008). Vigorous piglets that found the udder and suckled quickly had better survival rates. Sow characteristics are just as important: piglets were more likely to survive if the sow had a good placenta and if she showed good maternal behaviour. Ideally, sows should be calm and quiet during farrowing and they should lie down slowly and carefully thereby reducing the risk of crushing the piglets.

The potential for genetic selection was examined by studying gilts that were sired by boars from one of two genetic selection lines (High postnatal Survival (HS), or a Control Average Survival (C)), as well as their piglets in both indoor loose-housed and outdoor farrowing systems. Genotype affected total mortality at piglet level in the outdoor system but there was no effect in the indoor loose-housed environment. Genotype also influenced maternal characteristics: high survival gilts were more careful mothers in both environments, being significantly less likely to crush their piglets when changing posture during farrowing. The selection strategy had its greatest impact outdoors with 12% total mortality in the High Survival litters compared with 18% in the Control line.

In summary, these new selective breeding programmes for improved survival could not only benefit piglet and sow welfare but may also increase productivity and profitability for the farmer. The present findings also demonstrate the potential for phasing out the unpopular farrowing crates.

8.5 Alleviating social stress in beef cattle

Social stress caused by aggressive interactions or competition for resources such as food or lying space can be a major cause of poor welfare in many species and housing systems. As well as the deleterious effects of stress itself, aggressive interactions can cause injury, pain and even death. Furthermore, competition for food can disrupt the normal feeding pattern of cattle and, in turn, reduce their food intake and increase the risk of metabolic disturbances, such as rumen acidosis (Phillips and Rind, 2002). Subordinate animals show the greatest effects; they are more often displaced from the feeders, they shift their feeding patterns towards night time, they eat apart from dominants, and spend longer waiting around feeders to access the feed (Harb *et al.*, 1985; Hasegawa *et al.*, 1997; Ketelaar-de Lauwere *et al.*, 1996; McPhee *et al.*, 1964; Olofsson, 1999). A vicious circle may be established because increasing competition strengthens the relationship between social rank (or body weight) and feeding characteristics or feed intake (Collis, 1980; Friend *et al.*, 1977; Harb *et al.*, 1985; Katainen *et al.*, 2005; Olofsson, 1999).

It is thought that farmers may significantly underestimate the occurrence and importance of these undesirable consequences despite the fact that aggression and social stress can not only seriously damage the animals' welfare but they also reduce productivity, product quality, and therefore economic revenue. For example, if calves spend more days at the feedlot the costs of feed and management are higher while overall profit per head is lower. Encouragingly though, the level of social stress can be reduced by changes in housing conditions and feeding systems designed to reduce the need or motivation for animals to behave aggressively or to compete with each other for resources (see below).

8.5.1 Welfare Quality® research

In order to improve our understanding of the influence of social stress in intensively housed fattening cattle, Welfare Quality® researchers studied the effects of varying the number of animals per concentrate feeding place on performance, behaviour, selected welfare indicators, and rumen fermentation in feedlot heifers. Seventy-two Friesian heifers were used in a factorial arrangement with 3 treatments and 3 blocks of similar body weight. The treatments consisted of 2 (T2), 4 (T4), and 8 (T8) heifers per feeding place in the concentrate feeder (8 heifers/pen). Observations began after 4 weeks of adaptation to these treatments. Concentrate and straw were offered separately at 08:30 and the animals were fed *ad libitum*. During 6 periods of 28 days each, dry matter intake and average daily gain were measured, and blood and rumen samples were taken. The behaviour of the animals was also recorded.

The variability in final body weight between heifers sharing the same pen tended to rise and concentrate intake decreased linearly as competition increased. The proportions of abscessed livers increased quadratically with increased competition (8%, 4% and 20% in T2, T4 and T8 animals, respectively). The times spent eating concentrate decreased and eating rate increased linearly, whereas variability between pen-mates in concentrate eating time was greatest in T4 and T8. Increasing competition also resulted in a linear decrease in the time spent lying. The numbers of displacements from the concentrate feeders as well as the total sum of displacements increased linearly with increasing competition. The pen-average faecal corticosterone level was not affected by treatment but the maximum pen concentrations rose quadratically (greatest in T8), and dominant heifers were the most affected. The concentration of serum haptoglobin (which is an acute phase protein whose concentration is elevated when tissue damage occurs) increased linearly with competition, particularly in the most subordinate heifers. Increased competition reduced rumen pH in some of the experimental periods and increased rumen lactate; these effects are likely to lead to acidosis and ulceration.

In summary, the above results clearly suggest that increasing social pressure at the concentrate feeders beyond a threshold of 4 heifers per feeder has a negative effect on performance, health, product quality and animal welfare. Clear recommendations can therefore now be made.

8.6 Alleviating aggression and social stress in pigs

Aggression and social stress in pigs may be induced by a number of routine management practices and these undesirable events can seriously damage welfare and economic returns. Mixing unfamiliar animals together (often with a change of physical environment), is a common practice in pig husbandry, particularly at weaning and at the beginning of the growing-finishing period. This mixing can result in fighting as the pigs strive to establish dominance relationships, with most aggressive interactions being typically shown during the first few hours after grouping (Meese and Ewbank, 1972). Some studies reported reduced production after social mixing (e.g. Stookey and Gonyou, 1994), though others failed to do so (e.g. Coutellier et al., 2007). As stressors exert additive effects (Hyun et al., 1998), it is likely that the effects of social mixing will be more pronounced if it is combined with other stressors. This is almost certainly the case at weaning, when piglets are simultaneously subjected to nutritional, environmental and psychological stressors (Pluske et al., 1995); as a result they usually show a period of reduced feed intake that may have long-lasting effects on performance (Pollmann, 1993).

The frequency, duration and intensity of aggression after mixing can vary according to several factors, such as the time of day when the pigs are mixed, the amount of food provided and the presence or absence of environmental enrichment devices. For instance, weaned pigs offered tyres and hanging chains in the pen showed less aggression (Simonsen, 1990), and the provision of ad libitum food and regrouping after sunset were found to reduce the number of fights in the group (Barnett et al., 1994). The use of tranquillizing drugs to reduce aggression at mixing has also been widely advocated for many years, but although these can be helpful their efficacy becomes limited over time and agonistic interactions increase as the effects of the drugs wane (Gonyou et al., 1988). Furthermore, not only does this strategy simply treat the symptoms rather than the causes but the use of tranquillisers is also likely to lead to public concern. Group size can also affect how pigs react to being mixed with unfamiliar individuals, with larger groups showing less aggression. Encouragingly, the provision of certain pheromones (pig appeasing pheromone) can alleviate aggressive behaviour after regrouping (Guy et al., 2009; McGlone et al., 1987). Interestingly too, socialised piglets (those that were mixed with piglets from another litter before weaning) learned social skills that allowed them to more rapidly form stable hierarchies when regrouped after weaning (D'Eath, 2005).

Although some of the above treatments can exert beneficial effects the over-riding perception is that post-mixing aggression in commercial pig production is still a common phenomenon and that it cannot be significantly reduced by low-cost changes to the environment. On the other hand, a genetic component to individual aggressiveness has been described in pigs and many other species. Selective breeding against aggressiveness ought to be possible if a reliable and easily measured indicator trait can be shown to be genetically associated with aggressive behaviour. However, one must guard against the possibility that selection for reduced aggression might have undesirable effects if there are genetic correlations between aggressiveness and other characteristics such as the ease of handling, inactivity, decreased responsiveness, etc. Furthermore, observing individual differences in aggressiveness is time consuming and impractical for commercial breeders so a more easily measured criterion is needed.

8.6.1 Welfare Quality® research

Work done in Welfare Quality® had the following primary objectives:
- to estimate the genetic contribution to individual aggressiveness in pigs;
- to validate a method of predicting a pig's likely involvement in aggressive encounters based on a count of skin lesions (lesion score, LS) suffered following controlled mixing;
- to investigate genetic correlations between aggressive behaviour and other traits.

In order to estimate the genetic contribution to individual aggressiveness and to validate a quick and easy method of predicting involvement in aggressive encounters (based on LS scores), aggressive behaviour was recorded continuously for 24 h after pigs were placed in mixed groups and lesion scores were recorded at 24 h and 3 weeks post-mixing in 895 purebred Yorkshire pigs and 765 Yorkshire × Landrace pigs of both sexes. All the pigs were housed in partially slatted pens with straw bedding. Potential genetic correlations between aggressive behaviour and other traits were investigated by scoring behaviour during handling and general activity in the same population of 1,660 pigs.

Two behavioural traits were found to have a moderate to high heritability similar to that of growth traits; these were the duration of involvement in reciprocal fighting and the delivery of non-reciprocated aggression (NRA). On the other hand, receipt of NRA had a lower heritability. Genetic correlations suggested that the numbers of lesions to the anterior region of the body apparent at 24 h after mixing were associated with reciprocal fighting, receipt of NRA and, to a lesser extent, delivery of NRA. Lesions to the centre and rear of the body were associated primarily with receipt of NRA. Pigs which engaged in reciprocal fighting delivered NRA to other animals but rarely received NRA themselves; in other words pigs that fight also bully other animals but are rarely bullied themselves. Positive correlations were found between lesion scores observed 24 h and 3 weeks after mixing, especially for lesions to the centre and rear of the body. These important findings suggest that post-mixing lesions are predictive of those received under stable group conditions. Collectively, these results mean that aggressive pigs can be identified and selected against based on a distinctive pattern of skin lesions after mixing into a balanced group: the aggressive animals are genetically predisposed to have many lesions on their head and shoulders but fewer on other parts of the body (Turner *et al.*, 2009).

The researchers also measured the pigs' activity over a one-day period 3 weeks after mixing. Their behaviour was also measured when entering and exiting a weighing crate before mixing and again 3 months later just before reaching slaughter weight. Inactivity was weakly heritable and negatively associated with bullying, suggesting that pigs selected for reduced aggression might also be slightly less active. A greater diversity of scores and a higher heritability were found for the ease with which pigs entered a weigh crate than for the behaviour they showed in the crate or on exit. The ease with which the pigs entered and exited the crate had low positive genetic correlations with aggressive behaviours (fighting and bullying) but aggressive pigs were more active during weighing.

In summary, the findings show that selective breeding for reduced post-mixing lesion scores should have a long-term ameliorative effect on aggression and its related injuries even after dominance relationships have been established, i.e. the pigs will be generally less aggressive. A slight negative association between aggressiveness and activity might suggest that reducing aggression could make pigs less active and thereby harder to handle. However, it could also be argued that low aggression pigs were less stressed by handling and that they might also be less reactive to other stressful events such as transport and slaughter. More recently, it has been proposed that the identification of the causal functional molecular polymorphisms for aggressive behaviour and stress responsiveness (Murani *et al.*, 2010) could provide valuable markers for pig breeding.

8.7 Discussion and the way ahead

8.7.1 Improvement strategies and management support

As mentioned earlier (Chapter 4) feedback of the detailed results of the assessment measures to the farmer is a central part of the Welfare Quality® vision. An integral component of the feedback process is the provision of knowledge-based advice on remedial measures for welfare problems. Collectively, the Welfare Quality® efforts in this area focused on developing ways of minimising or eradicating the elicitation and expression of harmful behavioural and physiological traits and states in a range of farm animals, improving the human-animal relationship, and providing the animals with safe and stimulating environments. For instance, once the welfare status of a farm has been determined using the Welfare Quality® assessment protocols the feedback of results and the provision of practical advice (either from the assessor or an independent expert) on remedial strategies will help the farmer to deal successfully with any problems that were identified during the assessment.

In this context, some of the welfare improvement strategies developed in Welfare Quality® (such as the stockperson training programmes, welfare-friendly selection criteria for future breeding programmes, recommendations on housing and husbandry) will contribute significantly to the above advisory component of the cyclical process of farm assessment – feedback and advise – welfare improvement – reassessment. Moreover, by focusing on the 12 criteria and the welfare problems considered particularly important by a wide range of stakeholders, Welfare Quality® scientists are continuing to develop a Technical Information Resource (TIR) on practical welfare improvement strategies.

An advisory function is considered essential not only for on-going farm management, by helping the farmer to overcome problems, but also for the widespread and effective uptake of the welfare assessment protocols by farmers, auditors, advisors, retailers and other stakeholders. Furthermore, the implementation of viable welfare improvement strategies will support the development of new genotypes and innovative husbandry systems and practices that are designed to enhance the animals' quality of life, their productivity and product quality. Such effort will undoubtedly contribute to the diversification and societal sustainability of farm animal production in Europe and beyond.

8.7.2 Upgrading the Technical Information Resource on welfare improvement strategies

The remedial measures developed within and outside the Welfare Quality® project are described in the TIR mentioned above which also details the likely causes and consequences of welfare problems as well as potential risk factors (Jones and Manteca, 2009). Clearly there is an urgent need to maintain, update, extend (to include new strategies and new species) and disseminate (and/or offer easy access to) this information resource as new results emerge. A new collaborative venture that involves several of the Welfare Quality® partners as well as new participants (the Welfare Quality Network, www.welfarequality.net), may enable some of this important work to be carried out through additional and continuing support will undoubtedly be required.

Feedback from those involved in the continued development of the TIR and in the implementation and management of the new welfare-friendly systems and practices is considered very likely to highlight particular areas requiring further research. Such areas may include the fundamental biological knowledge required to underpin the assessment and improvement of: all aspects of animal welfare, relevant developments in stakeholder or consumer attitudes, possible areas of cross-compliance, and the informed formulation of social, economic and environmental policy. The prioritisation of such R&D needs would be extremely helpful for the decision-making process in a range of EU and national research funding bodies.

8.7.3 Welfare improvement strategies must be practical and effective

A number of critical requirements must be met before any strategy intended to improve farm animal welfare can be realistically implemented. Practicality is an overriding requirement. Thus, proposed 'welfare-friendly' changes to housing systems, to husbandry/management practices, or to breeding programmes must be shown to be practicable, robust, safe, affordable, easy to implement and in the long-term interests of the animals and the farmer. If these requirements are not met then the strategy will simply not be adopted.

It should also be recognised that, in the long term, an integrated approach involving the application of environmental, experiential, management and genetic strategies for welfare improvement is likely to be the most effective (Boissy *et al.*, 2005; Faure and Jones, 2004; Jones, 2004; Jones and Boissy, 2011; Jones *et al.*, 2004).

Genetic selection for or against specific behavioural and/or physiological traits is just one of the areas where particular care must be taken. It is important to ensure that selection for one characteristic is not unconsciously accompanied by the appearance of undesirable traits (Hocking *et al.*, 2011; Jones and Hocking, 1999). For example, while selection for good maternal behaviour is likely to improve welfare and production there is some concern that breeding against traits such as fear could result in generally unreactive animals or 'zombies' (D'Eath *et al.*, 2010), although the authors do concede that controls could be put in place to avoid such a negative outcome.

Despite the above cautionary notes, we must remember that several studies have pointed to the increasing likelihood of selection for welfare-friendly selection studies. For example, it has been shown that selective breeding of Japanese quail for reduced fear and adrenocortical stress responsiveness resulted in birds that were active and exploratory, with enhanced sociality, reduced developmental instability and increased productivity and product quality (Faure and Jones, 2004; Faure *et al.*, 2003; Jones and Hocking, 1999; Satterlee *et al.*, 2000). In other words, the low fear quail did not resemble 'zombies' at all. Similarly, low behavioural and physiological indices of stress are associated with enhanced activity, sociality and performance in dairy calves (Van Reenen *et al.*, 2005, 2009). Indeed, individual consistency of response patterns and the apparent durability of the relationships strongly suggest that there is scope for welfare-friendly and economically desirable breeding programmes in dairy cows (Van Reenen, personal communication).

Environmental enrichment represents another widely accepted way of enhancing the welfare status of domestic, zoo and farm animals. Enrichment generally involves increasing the stimulus value of the home environment; indeed the opportunity to investigate novel aspects of environments is thought to have particular value for all animals (Mench, 1994). Care is needed though because results can be inconsistent and many so-called enrichment stimuli have elicited very little or very short-lived interest (Jones, 2001, 2004, 2005). This probably reflected the fact that they were chosen according to human preconceptions of what the animal might find enriching. Research has slowly shifted towards 'asking the animal' what it finds interesting. Simplicity is another key feature but one which has often been underrated or totally neglected. In the real world the simpler the improvement strategy is then the more likely it is to be adopted. We will illustrate the above points with two examples from the poultry literature. Firstly, a systematic series of studies revealed that simple bunches of white baling twine were attractive to chickens of all ages (more so than chains, baubles or multi-coloured string), and they retained the birds' interest over very long periods (Jones, 2004, 2005; Jones et al, 2004). The birds drew the strings between their beaks and teased the strands apart; actions that strongly resembled preening. Furthermore, the incorporation of these simple string devices in the cages of laying hens reduced the expression of feather pecking as well as the amount of feather damage (Jones, 2004; Jones *et al.*, 2004; McAdie *et al.*, 2005). Bunches of string are now routinely used to minimise feather pecking and cannibalism in laying hens and broiler breeders housed in a variety of systems (Linda Keeling (SLU), Paul Cook (Farm Animal Initiative), Andrew Joret (Noble Foods), personal communication). Secondly, many farmers reported that playing the radio in their hen houses calmed the birds, reduced aggression, improved health and increased performance (Jones and Rayner, 1999). A possible interpretation is that the additional and varied sounds of the radio may help the birds learn that new sounds are not necessarily dangerous, thereby reducing fear and distress. Furthermore, switching on the radio is probably the easiest, most practical way of enriching the environment for the birds and the farmers.

8.7.4 Economic implications of welfare improvement

Costs incurred by the implementation of new methods of improving welfare, e.g. new breeding programmes, changes to husbandry systems, purchase of new equipment, attendance at training courses, etc., could conceivably deter some farmers. On the other hand, poor welfare is known to be associated with many costly consequences. Evidence is also gradually emerging for a positive effect of welfare improvement on productivity, product quality and profitability.

For example, the production losses, reduced product quality and related economic penalties of exposure to a variety of frightening stimuli, for example low flying aircraft, thunder, sudden and unfamiliar noises, have become increasingly apparent in a range of farm animals. In poultry for example, such events result in reduced growth rates, downgrading of carcasses at slaughter, eggshell abnormalities and poorer reproductive efficiency (Faure *et al.*, 2003; Jones, 1997). Similarly, claims by UK poultry producers for mortalities caused by low flying military aircraft cost the Ministry of Defence over £700,000 in 1995 (Jones, 1996). Although precise information on the economic consequences are not often available low fear of humans was associated with decreased egg production in laying hens (Hemsworth, 2003), better food conversion in broilers (Hemsworth *et al.*, 1994) and higher growth and reproductive performance in pigs (Hemsworth, 2003). It was even suggested that reducing broiler chickens' fear of humans could translate to a saving of AUS$ 8,400 per farm (Hemsworth, personal communication). Furthermore, stress-induced reduction in meat quality was less pronounced in a genetic line of Japanese quail showing low fear (Faure *et al.*, 2003), the incidence of pale soft exudative meat was lower in calm/docile pigs (Grandin and Deesing, 1998), and non-fearful cattle had less bruising and more tender meat (Fordyce *et al.*, 1988).

More recently, adaptive conjoint analysis (a market research technique) has been used to rank alternative management practices for extensive sheep farming systems in the United Kingdom by their perceived effect on animal welfare (Stott *et al.*, 2005). Farmers were asked to compare alternative policies defined by five attributes (labour, housing, veterinary treatment, feeding, and gathering) then the financial impact of these policies was assessed. Feeding attribute had the greatest positive impact on welfare while labour had a detrimental financial impact. Thus, effective labour management would be critical for sustainable extensive sheep farm systems. Because considerable variation in welfare score was observed at most farm income levels it was concluded that welfare could be improved within the context of viable farm business management by careful adoption of strategies to fit individual farm circumstances

A study designed to establish acceptable trade-offs between profit and welfare in alternative farrowing systems for pigs used linear programming (LP) to examine possible trade-offs and thereby aid the design of welfare-friendly but commercially viable alternatives (Ahmadi *et al.*, 2011). Crates yielded the highest annual net margin. However, there was scope for reducing piglet mortality, improving the sows' and piglets' welfare, and raising the net margin in the non-crate systems. Building on these results may lead to a win-win situation for the pigs and the farmers.

It would also appear that some of the improvement strategies developed in Welfare Quality® could have positive effects on productivity and profitability. For example, the implementation of certain of these strategies could lead to a reduction in the considerable losses in production due to health problems such as sub-acute acidosis in ruminants, lameness in broilers and dairy cattle, lesions caused by fighting after social mixing in pigs, and neonatal mortality in piglets.

Bennett and Thompson (2011) dissected the complexities of cost-benefit analysis while making a convincing case that economic analysis can play an important role in guiding decision makers concerning policy options and instruments designed to safeguard and improve farm animal welfare.

Despite the encouraging findings described above there remains a pressing need for an in-depth, industry-based investigation and cost-benefit analysis of the economic implications of welfare assessment and welfare improvement. Instead of a push based on legislation or the satisfaction of loans criteria a firm demonstration of the financial benefits of improved welfare would provide a far more potent incentive to farmers and other stakeholders to safeguard and enhance the wellbeing of their animals.

8.8 Conclusions

- The development, validation and dissemination of welfare improvement strategies underpins the advisory component of the cyclical process of farm assessment – feedback and advise – welfare improvement – reassessment.
- Improvement strategies may include 'welfare-friendly' changes to one or more of the following: housing and husbandry systems, stockmanship and management practices, and selective breeding programmes.
- Welfare Quality® researchers developed some potentially valuable improvement strategies including: a multi-media training programme for stockpersons; a feeding regime designed to reduce lameness in broiler chickens, genetic selection criteria to reduce aggression and neonatal mortality in pigs, a husbandry practice to lessen social stress in beef cattle, etc.
- A Technical Information Resource was developed which describes the causes and consequences of key welfare problems as well as practical solutions.
- A paramount requirement is that any new welfare improvement strategy must be practicable, safe, affordable, easy to implement, economically viable and in the long-term interests of the animals and the farmer, otherwise it will not be implemented.

- There is a pressing need for an industry-based cost/benefit analysis designed to establish the economic implications of welfare improvement and of specific improvement strategies.

Acknowledgements

Thanks are due to E. Baxter, R. D'Eath, A. Ferret. L. González, S. Jarvis. A. Lawrence, C. Leterrier and S. Turner for providing helpful information and references.

References

Ahmadi, V., Stott, A.W., Baxter, E.M., Lawrence, A.B. and Edwards, S.A. (2011). Animal welfare and economic optimisation of farrowing systems. Animal Welfare, 20, 57-67.

Barnett, J.L., Cronin, G.M., McCallum, T.H. and Newman, E.A. (1994). Effects of food and time of day on aggression when grouping unfamiliar adult pigs. Applied Animal Behaviour Science, 39, 339-347.

Baxter, E.M., Jarvis, S., D'Eath, R.B., Ross, D.W., Robson, S.K., Farish, M., Nevison, I.M., Lawrence, A.B. and Edwards, S.A. (2008). Investigating the behavioural and physiological indicators of neonatal survival in pigs. Theriogenology, 69, 773-778.

Bennett, R. and Thompson, P. (2011). Economics. In: Appleby, M.C., Mench, J.A., Olsson, I.A.S., Hughes, B.O. (eds.). Animal Welfare 2nd Edition. CAB International, Wallingford, UK, pp. 279-290.

Bizeray, D., Leterrier, C., Constantin, P., Picard, M. and Faure, J.M. (2000). Early locomotor behaviour in genetic stocks of chickens with different growth rates. Applied Animal Behaviour Science, 68, 231-242.

Boissy, A., Fisher, A.D., Bouix, J., Hinch, G.N. and Le Neindre, P. (2005). Genetics of fear in ruminant livestock. Livestock Production Science, 93, 23-32.

Boivin, X., Lensink, B.J., Tallet, C. and Veissier, I. (2003). Stockmanship and farm animal welfare. Animal Welfare, 12, 479-492.

Bradshaw, R.H., Kirkden, R.D. and Broom, D.M. (2002). A review of the aetiology and pathology of leg weakness in broilers in relation to welfare. Avian and Poultry Biology Reviews, 13, 45-103.

Collis, K.A. (1980). The effect of an automatic feed dispenser on the behaviour of lactating dairy cows. Applied Animal Ethology, 6, 139-147.

Corr, S.A., Maxwell, M.J., Gentle, D. and Bennett, D. (2003). Preliminary study of joint disease in poultry by the analysis of synovial fluid. Veterinary Record, 152, 549-554.

Coutellier, L., Arnould, C., Boissy, A., Orgeur, P., Prunier, A., Veissier, I. and Meunier-Salaun, M.C.(2007). Pig's response to repeated social regrouping and relocation during the growing-finishing period. Applied Animal Behaviour Science, 105, 102-114.

D'Eath, R. (2005). Socialising piglets before weaning improves social hierarchy formation when pigs are mixed post-weaning. Applied Animal Behaviour Science, 93, 199-211.

D'Eath, R.B., Conington, J., Lawrence, A.B., Olsson, I.A.S. and Sandøe, P. (2010). Breeding for behavioural change in farm animals: practical, economic and ethical considerations. Animal Welfare, 19, 17-27.

Dawkins, M.S., Donnelly, C.A. and Jones, T.A. (2004). Chicken welfare is influenced more by housing conditions than by stocking density. Nature, 427, 342-344.

Duncan, I.J.H. (1990). Reactions of poultry to human beings. In: Zayan, R. and Dantzer, R. (eds.) Social stress in domestic animals. Kluwer, Dordrecht, the Netherlands, pp. 121-131.

Edwards, S.A. (2002). Perinatal mortality in the pig: environmental or physiological solutions? Livestock Production Science, 78, 3-12.

European Commission (2000). The welfare of chickens kept for meat production. (Broilers). Report of the Scientific Committee on Animal Health and Animal Welfare, EC, Brussels, Belgium.

Faure, J.M. and Jones, R.B. (2004). Genetic influences on resource use, fear and sociality. In: Perry, G.C. (ed.). Welfare of the Laying Hen, 27th Poultry Science Symposium. CAB International, Wallingford, UK, pp. 99-108.

Faure, J.M., Bessei, W. and Jones, R.B. (2003). Direct selection for improvement of animal well-being. In: Muir, W. and Aggrey, S. (eds.) Poultry breeding and biotechnology. CAB International, Wallingford, UK, pp. 221-245.

FAWC (1992). Report on the Welfare of Broiler Chicken. Farm Animal Welfare Council, Surbiton, UK.

FAWC (1998). Report on the Welfare of Broiler Breeders. Farm Animal Welfare Council, Surbiton, UK.

Fordyce, G., Wythes, J.R., Shorthouse, W.R., Underwood, D.W. and Shepherd, R.K. (1988). Cattle temperament in extensive beef herds in northern Queensland 2. Effect of temperament on carcass and meat quality. Australian Journal of Experimental Agriculture, 28, 689-693.

Friend, T.H., Polan C.E. and McGilliard, M.L. (1977). Free stall and feed bunk requirements relative to behavior, production, and individual feed intake in dairy cows. Journal of Dairy Science, 60,108-118.

Gonyou, H.W., Parfet K.A., Anderson, D.B. and Olsson, R.D. (1988). Effects of Amperozide and Azaperone on aggression and productivity of growing-finishing pigs. Journal of Animal Science, 66, 2856-2864.

Grandin, T. and Deesing, M.J. (1998). Genetics and behavior during handling, restraint, and herding. In: Grandin, T. (ed.) Genetics and the behavior of domestic animals. Academic Press, San Diego, CA, USA, pp. 113-144.

Guy, J.H., Burns, S.E., Barker, J.M. and Edwards, S.A. (2009). Synthetic maternal pheromon affects post-mixing aggression and lesions in weaned pigs. Animal Welfare, 116, 45-51.

Harb, M.Y., Reynolds V.S. and Campling, R.C. (1985). Eating behavior, social dominance a voluntary intake of silage in group-fed milking cattle. Grass Forage Science, 40, 113-1

Hasegawa, N., Nishiwaki, A., Sugawara, K. and Ito, I. (1997). The effects of social exchange between two groups of lactating primiparous heifers on milk production, dominance order, behavior and adrenocortical response. Applied Animal Behaviour Science, 51,15-27.

Hemsworth, P.H. (2003). Human-animal interactions in livestock production. Applied Animal Behaviour Science, 81, 185-198.

Hemsworth, P.H. and Boivin, X. (2011). Human Contact. In: Appleby, M.C., Mench, J.A., Olsson, I.A.S., Hughes, B.O. (eds.) Animal welfare. 2nd Edition. CAB International, Wallingford, UK, pp. 246-262.

Hemsworth, P.H. and Coleman, G.J. (1998). Human-Livestock Interactions: the Stockperson and the Productivity of Intensively Farmed Animals. CAB International, Wallingford, UK.

Hemsworth, P.H., Coleman, G.J., Barnett, J.L. and Jones, R.B. (1994). Behavioural responses to humans and the productivity of commercial broiler chickens. Applied Animal Behaviour Science, 41, 101-104.

Herpin, P., Damon, M. and LeDividich, J. (2002). Development of thermoregulation and neonatal survival in pigs. Livestock Production Science, 78, 25-45.

Hocking, P.H., D'Eath, R.B. and Kjaer, J.B. (2011). Genetic Selection. In: Appleby, M.C., Mench, J.A., Olsson, I.A.S., Hughes, B.O. (eds.) Animal welfare, 2nd Edition. CAB International, Wallingford, UK. pp. 263-278.

Holyoake, P.K., Dial, G.D., Trigg, T. and King, V.L. (1995). Reducing pig mortality through supervision during the perinatal period. Journal of Animal Science, 80, 68-78.

Hyun, Y., Ellis, M. and Johnson, R.W. (1998). Effect of feeder type, space allowance, and mixing on the growth performance and feed intake pattern of growing pigs. Journal of Animal Science, 76, 2771-2778.

Jones, R.B. (1993). Reduction of the domestic chick's fear of humans by regular handling and related treatments. Animal Behaviour, 46, 991-998.

Jones, R.B. (1995). Habituation to human beings via visual contact in docile and flighty strains of domestic chicks. International Journal of Comparative Psychology, 8, 88-98.

..es, R.B. (1996). Fear and adaptability in poultry: insights, implications and imperatives. ..orlds' Poultry Science Journal, 52, 131-174.

..B. (1997). Fear and distress. In: Appleby, M.C. and Hughes, B.O. (eds.). Animal CAB International, Wallingford, UK, pp. 75-87.

.(1998). Alleviating fear in poultry. In: Greenberg, G. and Haraway, M. (eds.) ..ve psychology: a handbook. Garland Press, New York, NY, USA, pp. 339-347.

..01). Environmental enrichment for poultry welfare. In: Wathes, C.M. (ed.) ..anagement systems for livestock. British Society for Animal Science, ..blication, No 28. Pp. 125-131.

Environmental enrichment: the need for practical strategies to improve ..In: Perry, G.C. (ed.). Welfare of the laying hen. CAB International, ..p. 215-225.

..onmental enrichment can reduce feather pecking In: Glatz, P. (ed.). ..rsity of Nottingham Press, Nottingham, UK, pp. 97-100.

Jones, R.B. and Boissy, A. (2011). Fear and other negative emotions. In: Appleby, M.C., Mench, J.A., Olsson, I.A.S., Hughes, B.O. (eds.) Animal welfare 2nd Edition. CAB International, Wallingford, UK, pp. 78-97.

Jones, R.B. and Hocking, P.M. (1999). Genetic selection for poultry behaviour: big bad wolf or friend in need? Animal Welfare, 8, 343-359.

Jones, R.B. and Manteca, X. (2009). Best of Breed. Public Science Review, 18, 562-563.

Jones, R.B. and Rayner, S. (1999). Music in the hen house: a survey of its incidence and perceived benefits. Poultry Science, 78, 110.

Jones, R.B., Blokhuis, H.J., De Jong, I.C., Keeling, L.J., McAdie, T.M. and Preisinger, R. (2004). Feather pecking in poultry: the application of science in a search for practical solutions. Animal Welfare, 13, S215-219.

Julian, R. (2005). Production and growth related disorders and other metabolic diseases of poultry – a review. Veterinary Journal, 169, 350-369.

Julian, R.J. (1998). Rapid growth problems: ascites and skeletal deformities in broilers. Poultry Science, 77, 1773-1780.

Katainen, A., Norring, M., Manninen, E., Laine, J., Orava, T., Kuoppala, K. and Saloniemi, H. (2005). Competitive behaviour of dairy cows at a concentrate self-feeder. Acta Agricultura Scandinavica Section A, 55, 98-105.

Kestin, S.C., Adams, J.M. and Gregory, N.G. (1994). Leg weakness in broilers, a review of studies using gait scoring. European Poultry Conference, pp. 203-206.

Kestin, S.C., Gordon, S., Su, G. and Sorensen, P. (2001). Relationships in broiler chickens between lameness, live weight, growth rate and age. Veterinary Record, 148, 195-197.

Kestin, S.C., Knowles, T.G., Tinch, A.E. and Gregory, N.G. (1992). Prevalence of leg weakness in broiler chickens and its relationship with genotype. Veterinary Record, 131, 190-194.

Kestin, S.C., Su, G. and Sorensen, P. (1999). Different commercial broiler crosses have different susceptibilities to leg weakness. Poultry Science, 78, 1085-1090.

Ketelaar-de Lauwere, C.C., Devir, S. and Metz, J.H.M. (1996). The influence of social hierarchy on the time budget of cows and their visits to an automatic milking system. Applied Animal Behaviour Science, 49, 199-211.

Knol, E.F., Leenhouwers, J.I. and Van der Lende, T. (2002). Genetic aspects of piglet survival. Livestock Production Science, 78, 47-55.

Malmkvist, J., Pedersen, L.J., Damgaard, B.M., Thodberg, K., Jorgensen, E. and Labouriau, R. (2006). Does floor heating around parturition affect the vitality of piglets born to loose housed sows? Applied Animal Behaviour Science, 99, 88-105.

McAdie, T.M., Keeling, L.J., Blokhuis, H.J. and Jones, R.B. (2005). Reduction in feather pecking and improvement of feather condition with the presentation of a string device to chickens. Applied Animal Behaviour Science, 93, 67-80.

McGlone, J.J., Curtis, S.E. and Banks, E.M. (1987). Evidence for aggression-modulating pheromones in prepuberal pigs. Behavioural and Neural Biology, 47, 27-39.

McNamee, P.T., McCullagh, J.J., Rodgers, J.D., Thorp, B.H., Ball, H.J., Connor, T.J., McConaghy, D. and Smyth, J.A. (1999). Development of an experimental model of bacterial chondronecrosis with osteomyelitis in broilers following exposure to *Staphylococcus aureus* by aerosol, and inoculation with chicken anaemia and infectious bursal disease viruses. Avian Pathology, 28, 26-35.

McPhee, C.P., McBride, G. and James, J.W. (1964). Social behavior of domestic animals. III. Steers in small yards. Animal Production, 6, 9-15.

Meat and Livestock Commission (2006). The pig yearbook, MLC, Milton Keynes, UK.

Meese, G.B. and Ewank, R. (1972). A note on instability of the dominance hierarchy and variations in level of aggression within groups of fattening pigs. Animal Production, 14, 359-362.

Mench, J.A. (1994). Environmental enrichment and exploration. Laboratory Animal, 23, 38-41.

Morrison, V., English, P.R. and Lodge, G.A. (1983). The effect of alternative creep heating arrangements at two house temperatures on piglet lying behaviour and mortality in the neonatal period. Animal Production, 36, 530-531.

Mülleder, C. and Waiblinger, S., 2004. Analyse der Einflussfaktoren auf Tiergerechtheit, Tiergesundheit und Leistung von Milchkühen im Boxenlaufstall auf konventionellen und biologischen Betriebe unter besonderer Berücksichtigung der Mensch-Tier-Beziehung. Endbericht FP 1264. VUW, Wien, Austria, 165pp.

Muráni, E., Ponsuksili, S., D'Eath, R.B., Turner, S.P., Kurt, E., Evans, G., Thölking, L., Klont, R., Foury, A., Mormède, P. and Wimmers, K. (2010). Association of HPA axis-related genetic variation with stress reactivity and aggressive behaviour in pigs. BMC Genetics, 11, 74-85.

Olofsson, J. (1999). Competition for total mixed diets fed for ad libitum intake using one or four cows per feeding station. Journal of Dairy Science, 82, 69-79.

Phillips, C.J.C. and Rind, M.I. (2002). The effects of social dominance on the production and behavior of grazing dairy cows offered forage supplements. Journal of Dairy Science, 85, 51-59.

Pickup, H.E., Cassidy, A.M., Danbury, T.C., Weeks, C.A., Waterman, A.E. and Kestin, S.C. (1997). Self selection of an analgesic by broiler chickens. British Poultry Science, 38, S12-S13.

Pluske, J.R., Williams, I.H. and Aherne, F.X. (1995). Nutrition of the neonatal pig. In: Varley, M.A. (ed.) The neonatal pig. development and survival. CAB International, Wallingford, UK, pp. 187-235.

Pollmann, D.S. (1993). Effects of nursery feeding programs on subsequent grower-finisher pig performance. In: Martin, J. (ed.) Proceedings of the Fourteenth Western Nutrition Conference. Faculty of Extension, University of Alberta, Edmonton, Canada. pp. 243-254.

Sanotra, G.S. (2000). Leg problems in broilers: A survey of conventional production systems in Denmark. Dyrenes Beskyttelse, Frederiksberg, Denmark.

Sanotra, G.S., Berg, C. and Lund, J.D. (2003). A comparison between leg problems in Danish and Swedish broiler production. Animal Welfare, 12, 677-683.

Sanotra, G.S., Lund, J.D., Ersboll, A.K., Petersen, J.S. and Vestergaard, K.S. (2001). Monitoring leg problems in broilers: a survey of commercial broiler production in Denmark. World's Poultry Science Journal, 57, 55-69.

Satterlee, D.G., Cadd, G.G. and Jones, R.B. (2000). Developmental instability in Japanese quail genetically selected for contrasting adrenocortical responsiveness. Poultry Science, 79, 1710-1714.

Seabrook, M.F. and Bartle, N.C. (1992). Human factors. In: Phillips, C. and Piggins, D. (eds.) Farm animals and the environment. CAB International, Wallingford, UK, pp. 111-125.

Simonsen, H.B. (1990). Behaviour and distribution of fattening pigs in the multi-activity pen. Applied Animal Behaviour Science, 27, 311-324.

Sorensen, P., Su, G. and Kestin, S.C. (2000). Effects of age and stocking density on leg weakness in broiler chickens. Poultry Science, 79, 864-870.

Stookey, J.M. and Gonyou, H.W. (1994). The effects of regrouping on behaviour and production parameters in finishing pigs. Journal of Animal Science, 72, 2804-2811.

Stott, A.W., Milne, C.E, C.E., Goddard, P.J. and Waterhouse, A. (2005). Projected effect of alternative management strategies on profit and animal welfare in extensive sheep production systems in Great Britain. Livestock Production Science, 97, 161-171.

Tuchscherer, M., Puppe, B., Tuchscherer, A. and Tiemann, U. (2000). Early identification of neonates at risk: traits of newborn piglets with respect to survival. Theriogenology, 54, 371-388.

Turner, S.P., Roehe, R., D'Eath, R.B., Ison, S.H., Farish, M., Jack, M.C., Lundeheim, N., Rydhmer, L. and Lawrence, A.B. (2009). Genetic validation of postmixing skin injuries in pigs as an indicator of aggressiveness and the relationship with injuries under more stable social condition. Journal of Animal Science, 87, 3076-3082.

Van Reenen, C.G., Hopster, H., van der Werf, J.T.N., Engel, B., Buist, W.G., Jones, R.B. and Blokhuis, H.J. (2009). The benzodiazpeine brotizolam reduces fear in calves exposed to a novel object test. Physiology and Behavior, 96, 307-314.

Van Reenen, C.G., O'Connell, N.E., Van der Werf, J.T.N., Korte. S.M., Hopster, H., Jones, R.B. and Blokhuis, H.J. (2005). Responses of calves to acute stress: Individual consistency and relations between behavioral and physiological measures. Physiology and Behavior, 85, 557-570.

Waiblinger, S. and Spoolder, H. (2007). Quality of stockpersonship. In: Velarde, A. and Geers, R. (eds.) On-farm monitoring of pig welfare. Wageningen Academic Publishers, Wageningen, the Netherlands, pp. 159-166.

Waiblinger, S., Boivin, X., Pedersen, V., Tosi, M., Janczak, A.M., Visser, E.K., Jones, R.B. (2006). Assessing the human-animal relationship in farmed species: a critical review. Applied Animal behaviour Science, 101, 185-242.

Wathes, C. and Whittemore, C.T. (2006). Environmental management of pigs. In: Kyriazakis, I. and Whittemore, C.T. (eds.). Science and practice of pig production. Blackwell Publishing, Oxford, UK, pp. 533-592.

Weeks, C.A., Danbury, T.D., Davies, H.C., Hunt, P. and Kestin, S.C. (2000). The behaviour of broiler chickens and its modification by lameness. Applied Animal Behaviour Science, 67, 111-125.

White, K.R., Anderson, D.M. and Bate, LA. (1996). Increasing piglet survival through an improved farrowing management protocol. Canadian Journal of Animal Science, 76, 491-495.

Chapter 9. Relevance and implementation of Welfare Quality® assessment systems

Andy Butterworth, Harry Blokhuis, Bryan Jones and Isabelle Veissier

9.1 Introduction

As indicated in earlier chapters, there is widespread interest in the possible future uptake and implementation of a welfare assessment system focusing mainly on animal-based measures. For example, such a system could help farmers to improve welfare and production and to minimise disease risks by providing them with information on 'how they are doing' and on 'what areas require attention'. The system could also assist the efforts of farmers and farmer groups to add value through improved welfare and to make associated claims. As well as improving the animals' quality of life a reliable, trustworthy welfare assessment system could also benefit farm management, farm assurance and certification, the provision of consumer information, veterinary inspection of farms, and retailer claims concerning high quality food products. It could also help legislators to move to a more outcome based type of legislation regarding animal welfare. Additionally, scientific researchers may wish to use an 'agreed' assessment framework in their farm research.

Regardless of the end user, the application of the Welfare Quality® assessment protocols involves a number of steps. Clearly, the potential user(s) should first be provided with the appropriate Welfare Quality® protocol which contains detailed and standardised descriptions of the measures and a practical guide to carrying them out (these include animal-, resource- and management-based measures). The next step involves the provision of thorough training in the methods followed by an evaluation of the trainees' ability to apply the protocol in a correct, uniform and repeatable manner. The application of the protocol in practice enables the determination of the welfare status of the animal unit.

With an aim to safeguard and improve animal welfare the results of this assessment should not only be fed back to the manager of the animal unit but should also be followed by an identification of areas requiring attention, the identification of related risk factors and the implementation of appropriate remedial measures.

Many people from different backgrounds and from different types of organisation have already been trained by former partners of the Welfare Quality® project in the use of the Welfare Quality® assessment protocols. These 'trained' individuals now apply this animal assessment knowledge in their professional environment and have started to use animal-based assessment for a variety of purposes. In this chapter we describe and discuss a possible 'road map' that could help those wishing to adopt Welfare Quality® assessment protocols in supply chains or food assurance schemes to follow an established track. We also discuss numerous aspects of training and what lessons we can learn from the outcomes of early implementation activities. Finally we present some thoughts on the applicability and value of Welfare Quality® methods for a variety of purposes (e.g. cross compliance, food labelling, setting welfare targets, monitoring progress, assessing innovative farming systems/new breeds).

9.2 A stepwise approach (road map) to using the Welfare Quality® assessment system

Before the Welfare Quality® assessment system can be adopted in commercial assurance programmes an agreed process must be put in place. This agreement will ensure that any organisation that plans to use the Welfare Quality® assessment protocols will have examined and decided how it will actually use the methods in practice, who will carry out assessments, how these future assessors will be trained and validated, how the results of the assessments will be used, how often assessments should be carried out, how claims of compliance with the Welfare Quality® assessment system will be described in advertising and product claims, and how the quality and robustness of the process will be maintained over time. To facilitate the fulfilment of these requirements partners of the Welfare Quality Network (WQN) (see Chapter 10) recently prepared a stepwise 'road map' which describes the progressive steps or phases involved in the uptake of the Welfare Quality® assessment system. The road map also details the claims that the end user can make upon completion of each step; this will help ensure that companies make only realistic and justifiable claims as to the degree of incorporation of Welfare Quality® methods into their welfare assessment schemes. The phases of the road map stages are:

9.2.1 Phase 1: scope and early fact finding

In this phase a company may approach a WQN partner because it plans to incorporate Welfare Quality® measures and protocols in its commercial assessment regime. The Welfare Quality Network will then work together with the company to help define the goals and the most effective ways of implementing the protocols. This may include, for instance, determining the likely number of farms involved, the persons(s) who will carry out the assessments, the aspiration level (acceptable, enhanced, excellent), whether all the company's farms will be included and whether the aspiration level will be determined by the farm with the lowest welfare score/level, the expected time path, and how the outcomes will be used (business to business, product marketing, use of an 'own label', use of the Welfare Quality® logo, etc.). Thus, this first phase focuses primarily on exploring the needs of the proposed users and the best ways of satisfying these needs

9.2.2 Phase 2: fact finding

The WQN partner who is directly involved with the company visits a sample of the farms that will be included in the assessment programme and carries out trial Welfare Quality® assessments. The aim is to get an idea of the welfare status of those farms and to estimate the scope for improvement. The capacities of the company (organisation, infrastructure) relevant for a successful implementation of WQ assessments are also evaluated to help establish which approaches will best suit the company and its training needs.

9.2.3 Phase 3: define implementation route

The WQN partner, perhaps in collaboration with other Network partners, will help to define the most suitable programme of activities to achieve the goal(s) defined in the above phases. This exercise includes defining the requirements for training, the necessary numbers of assessors and farm visits, data handling, the precise involvement of WQN partners, etc. A detailed description of a pilot phase and a full time path will also be formulated.

9.2.4 Phase 4: pilot implementation

In this pilot activity, assessors linked to the company are trained by WQN partner(s) or by other trainers certified in the use of the appropriate Welfare Quality® protocols to carry out welfare assessments at a selection of farms in the company's supply chain. The data are then handled and integrative scores are calculated by WQN partners to provide an overall welfare score for that farm or animal unit. The outcomes of this pilot implementation will enable the company to decide how the Welfare Quality® scoring system can (or cannot) be used to meet its needs.

9.2.5 Phase 5: decision on full implementation

On the basis of discussion of the data from the pilot implementation (phase 4), steps will be taken to advance toward full implementation of the Welfare Quality® system. If there is agreement to continue, the commercial partner and the WQN partner most directly involved will formulate a plan for implementation of Welfare Quality® assessments in the whole supply chain as well as related requirements (e.g. changes of routines, farmer involvement) and aspiration (e.g. improvement strategies). This plan will then need to be agreed between the Welfare Quality Network management team, the involved WQN partner(s) and the company.

9.2.6 Phase 6: full implementation

Once full agreement has been reached the company will be permitted to roll-out the implementation plan to its farms; this plan may include further training, additional assessments, calculation of scores (by WQN), the effective communication of results and identification of the target audience (consumers, other businesses, or other stakeholders). The implementation programme may also include the use of the Welfare Quality® logo to identify (and publicise) the welfare level achieved (acceptable, enhanced, excellent).

The final outcome of this step wise process would be a fully implemented assessment scheme including a growing database of results. The commercial partner could then make labelling and marketing claims: for the assessed farms: 'Welfare Quality® assessed', for the company 'Our products are Welfare Quality® assessed'

9.3 Training programmes and lessons learned

9.3.1 Different needs for different users

Sets of training activities to support the application of the Welfare Quality® protocols for pigs, poultry (broilers and layers) and cattle (dairy and beef) have already been developed. Early in the project we recognised that different groups of end-users are likely to have different training needs (some of these are listed below). Therefore, the creation of practical and agreed training to suit the needs of different users has been an on-going effort since the delivery of the WQ® protocols. The aim was to create 'harmonised' training, materials and approaches across the species (cattle, pigs, poultry) such that an assessor can be trained to reliably carry out the required WQ® protocols and assessment methods. He/she should then be 'examined' so that there is assurance that his/her assessment performance will be credible, repeatable and consistent. The training materials and methods in use today originated from the learning and resource material which was first used to train the technical and research staff in the institutes carrying out the projects within the WQ® project. The durations of current training programmes carried out by partners in the Welfare Quality Network (see Chapter 10) for different animal categories vary from 1 day classroom activity and 1 day on farm for pigs, broilers and layers to 1 day classroom and 2-2.5 days on farm for cattle. The courses utilise video material and photographic reference material. A practical example focusing on one measure (bird cleanliness' in broiler chickens) is shown in Box 1.

Box 1. A practical example of how Welfare Quality® protocols can be taught and then used in practice – 'bird cleanliness'.

In this example we focus on one measure 'cleanliness in broiler chickens' as assessed on farm. Feathers keep birds warm and dry, prevent them becoming dirty, and reduce the risk of skin infections. Birds spend much time preening their feathers and this helps them stay clean and healthy. If the feathers become wet or soiled with litter (bedding), faeces or dirt they can lose their protective properties; this can harm the birds' welfare. The Welfare Quality® scoring scale enables reliable assessment of the degree of cleanliness in the birds as well as an effective method of differentiating flocks and farms (much greater numbers of dirty birds are found in some bird units than others).

Training programme and measures

In a training programme assessors are made familiar with the protocols and measures via one day of classroom training and one day carrying out an on farm assessment (inter-observer reliability testing is also included here). Trainees then return to the classroom for feedback and reporting. The overall exercise ensures their familiarity with the practical details and the time required.

More specifically, training in scoring for cleanliness involves examination of photographs and practice during the on-farm visit. The trainees are assessed during the training course until they are capable of a uniform scoring, and will be asked to carry out periodic validation or reference audit to check that they continue to score in a repeatable way.

The assessment

One hundred birds are assessed, 10 from each of 10 areas of the house, including 2 areas near the drinkers, 2 near to feeders, 3 near a wall, and 3 in the resting area. Individual birds are picked up carefully, scored for cleanliness and returned to the flock.

Background information is obtained via a farm questionnaire or standard inspection report. This includes: (1) broiler breeder information including genotype/strain, history and age; (2) hatchery information, e.g. distance/time transported, chick vaccination programme; (3) general information such as number and weight of chicks placed, sex, time of year, age at assessment and slaughter; (4) specific husbandry practices and housing like stocking density and thinning, brooding conditions, nutritional profile, vitamin/mineral levels, litter substrate, feeder and drinker design/type, lighting programme, age of house, construction details, target ventilation profile, diseases and medication history, vaccination programme and water source; (5) performance and health data e.g. growth profile, weekly patterns of mortality, leg culls and other culls; and (6) background information about the site/company including numbers of birds on site and biosecurity measures. Collectively this information can help identify potential risk factors and thereby guide the farmer on how to resolve the welfare issues.

>>>

<<<

Analysis of risk factors and remedial measures

The farm, company or assurance/advisory service can analyse the information and determine the prevalence and severity of dirty birds in the flock. Subsequent welfare improvements may for instance involve resolving poor litter quality (e.g. due to inadequate ventilation, inappropriate nutritional factors or leaking drinkers). In some countries, inspection bodies focus on litter quality as a marker for company welfare performance, with bird cleanliness, foot pad lesions and hock burn considered as proxy indicators.

Comparisons between 'good' and 'poor' farms can also help identify risk factors associated with differences in management, environment, feeding, medication, stockmanship or genetics.

There is clearly a need for continued development of training and examination practices. Indeed, since the WQ® project finished in December 2009 many organisations have requested training in the WQ® protocols. The needs of these organisations fall into several distinct categories:

- Government inspection bodies (e.g. state veterinary staff) requiring training in the protocols as a part of their commitment to general awareness of ways to assess animal welfare on farm and at slaughter.
- Research institutions (i.e. scientists) wishing to use standardised assessment methods for research purposes.
- Animal Welfare NGO's – who recognise that training in practical welfare assessment is relevant to the needs of their staff and the organisations that they work with.
- Farm assurance companies – who consider the inclusion of animal-based measures in their assurance and certification work.

9.3.2 Insights gained from training different types of users

Training in the use of Welfare Quality® assessment protocols has now taken place in numerous countries across the EU, Eastern Europe, Asia and Latin America. The main conclusions are:

- As it stands, the training programme has been very travel intensive. Not only has it been necessary to visit whichever country the contracting organisation is based in but also to travel within that country from farm to farm to conduct face-to-face training. It may be more feasible to streamline training activities by centralising the location and carrying out 'joint training events' to which end users from a number of organisations and countries are invited.
- It is conceivable that a generalised 'introduction to animal welfare' module may not be needed for all trainees; its removal could certainly shorten the process.

However, many trainees feel that this section is essential. Currently, there seem to be two main types of trainee:

▷ Type 1 – Trainees who are not interested in, or do not want, assessment training to the level required for certification purposes, but who seem more interested in learning about general concepts of animal welfare assessment rather than in actually carrying out an assessment. This type includes, for example, veterinary groups, state veterinary organisations, legislators and ministry officials.

▷ Type 2 – Those requiring training for 'higher level/certification level' purposes fall into two main groups: (a) scientific researchers who wish to use the assessment protocols in specific projects where accurate, repeatable and consistent use of the measures is essential; and (b) assessors from certification bodies who are contemplating the use of outcome based measures in farm inspection schemes. Both of the Type 2 groups require not only classroom and farm training, but also the full inter observer and repeatability exercises.

- For those organisations/individuals who undergo 'higher level' training, it will be necessary to establish a 'pass/fail' decision system so that only those candidates who achieve an agreed level of performance are approved to carry out the protocols for certification purposes. Thus, a standardised examination with an agreed pass/fail threshold is needed.

- Organisations that wish to make reference to the Welfare Quality® system or to use the whole (or part) of it in their certification processes should be required to use trained/approved assessors. On-going monitoring of approved assessors is necessary to ensure that they maintain an appropriate level of consistency.

- The option of creating 'trained trainers' should be considered so that training courses can be made more widely available. Thus, existing groups of species-specific trainers could train two or more other trainers to certification level, partly to ensure that these training skills are not lost as well as to ensure that the 'system' has the capacity to respond to additional demand.

- The Welfare Quality® assessment training course could arguably be developed into a distance learning/E-Learning scheme. However, whilst it may be possible to create an electronic learning module for the introductory concepts of outcome based welfare assessment, it is also clear that practical and 'on farm' training is best carried out face to face, and that the final 'approval' of the ability to carry out the assessment in a reliable and repeatable way should be practically demonstrated/examined. Such 'quality control' is considered critical, particularly during the early stages, to ensure credibility and confidence in the system.

- Compared to the costs of commercial training courses, the relatively low cost of Welfare Quality® assessment training events represent particularly good value. Although many potential clients may expect such training to be inexpensive, decisions on realistic pricing need to be made and communicated.

9.4 Some issues of applicability

The Welfare Quality® assessment system aimed to be applicable to a wide range of situations. Indeed, the 12 welfare criteria are in theory applicable to all species of terrestrial vertebrates. There might nevertheless be some criteria which are less pertinent in some circumstances. For instance a good expression of social or other behaviours (Criteria 9 and 10) do not seem critical at slaughter. Therefore these criteria are not included in the protocol for assessing the welfare of animals at slaughter.

Even though a particular criterion may be relevant for a given animal type there are not always adequate measures available. Assessing the thermal comfort of adult cattle is a case in point; animal-based measures that might detect discomfort due to heat (increased respiratory rate, decreased feeding behaviour) are neither sufficiently sensitive nor feasible on farm and responses to cold are not clearly visible. At the same time, resource-based measures (such as temperature, humidity, air speed) are difficult for a non-expert person to apply. Therefore, there is currently no measure of thermal comfort in the Welfare Quality® cattle protocol.

The ways in which the various criteria are checked differ between animal types and farming systems. For instance, lameness is detected by observing barn-housed cows walking whereas for tethered animals the observer seeks to identify animals that do not bear their body weight on all feet or that shift the weight between feet.

Whilst the current Welfare Quality® protocols have been developed in a way that they can be applied in several farming systems they do not yet cover all types. For example, the pig and cattle protocols designed for animals that are housed indoors need to be adapted to suit the assessment of animals in outdoor systems.

9.5 Implementing the Welfare Quality® assessment system for different purposes

9.5.1 Providing standards for cross-compliance

To check cross compliance of farms or slaughter plants with the relevant EU welfare legislation and to determine that they meet certain minimum welfare standards in agreement with the general EU directive to protect farm animal welfare (European Commission, 1998), it is necessary to distinguish those animal units which are above from those which fall below the minimum legal requirements or equivalent. When a specific legislation exists, e.g. for hens (European Commission, 1999), for pigs (European Commission, 2001a,b), for broilers (European Commission, 2007), and calves (European Commission, 2008), minimal norms are often set. For instance, the maximum stocking density for broilers is 33 kg/m^2, sows and gilts shall have permanent access to materials for rooting, calves should not be housed in individual crates after 8 weeks of age and the crates should be at least as wide as the calf height plus 10%. In these cases, the compliance with the legislation is derived from whether or not the farm satisfies these specific rules (RBMs).

In other welfare directives, such norms are replaced by the need to attain a certain level of results. For instance, according to Council regulation 1099 (European Commission, 2009b) for the protection of animals at slaughter it is necessary to check that animals are stunned effectively, i.e. that they are not conscious at the time they are killed. The regulation does not describe how consciousness should be evaluated. In this case the measures proposed in the Welfare Quality® assessment for slaughter can be used. Similarly in the directive for broilers (European Commission, 2007), the minimal space allowance of 33 kg/m^2 can be increased up to 39 kg/m^2 if stricter welfare standards, e.g. regarding the prevalence of pododermatitis, are taken into account. Here too the Welfare Quality® measure of pododermatitis can be used to determine compliance.

In the event that no specific directive exists but the welfare status of animals still needs to be determined (e.g. for adult cattle) it is necessary to use an ad-hoc tool to check the compliance of farms with European welfare standards. The Welfare Quality® assessment system is suitable for such a use. Indeed the lowest category ('not classified') corresponds to farms that are not acceptable, since the category immediately above corresponds to 'acceptable' or 'above minimal requirements' (see Chapter 7). An animal unit (farm or slaughter plant) falls in the 'not classified' category if it scores 20 or below for two or more of the welfare principles (on the 0-100 values scale defined in Welfare Quality®), where a score of 20 or above corresponds to attainment of legislative requirements or their equivalents. It is therefore proposed that the Welfare Quality® assessment system could be used to check the cross compliance of farms, where 'not classified' farms could be considered as 'not complying' with EU welfare standards. Of course, this proposal needs to be agreed with and adopted by legislative bodies.

9.5.2 Labelling levels of welfare

The European Economic and Social Committee concluded that a labelling system based on scientific evidence was necessary to identify products with higher welfare standards (Polten, 2007). In 2009, the European Commission re-examined the 'possibility of establishing a system of animal welfare labelling to improve consumer information on welfare standards and existing welfare schemes and to harmonise the internal market to prevent widely differing welfare standards being used under the generic 'welfare' term' (European Commission, 2009a). Several options for labelling were explored by the Commission. These include mandatory vs. voluntary labelling:

Mandatory labelling systems

The first option investigated for mandatory labelling corresponds to labelling of the welfare standards under which animal products are produced. The Welfare Quality® assessment system offers the possibility to distinguish animal units that attain welfare standard above legal requirements (or equivalents) from those that merely comply with these requirements. Indeed, the Welfare Quality® assessment system defines two additional welfare categories ('enhanced' and 'excellent') which require higher welfare scores than those in the 'acceptable' category which simply corresponds to legislative requirements or their equivalents.

Another mandatory labelling system explored by the Commission is linked to the farming system itself, i.e. products are labelled if they have been produced in a system that is recognised as providing good welfare. If such a labelling system is put in place it will require assessment of the levels of welfare provided by the main production systems. Once again the Welfare Quality® assessment system could be effectively applied to satisfy such requirements.

Voluntary labelling systems

The voluntary labelling options considered by the EC include:
- establishment of requirements for voluntary use of animal welfare claims, i.e. what terms could be used on a label;
- establishment of a voluntary Community Animal Welfare Label open for all to use if they meet the criteria, i.e. the assessment and animal welfare requirements that must be met to use the defined labelling terms;
- drafting of guidelines for animal welfare labelling and quality schemes, i.e. the administrative procedures and protocols necessary to operate an assurance and product quality scheme in line with established requirements (European Commission, 2009a).

An independent third party inspection is likely to be recommended in order to ensure credibility. An information system linking categories of welfare to specific threshold scores may also be required here. Again the Welfare Quality® assessment system can be used, with a specific focus on the categories 'enhanced' and 'excellent'.

Citizens generally consider animal welfare within a bundle of other issues related to animal products, including quality, environment preservation, and human health

(Miele *et al.*, 2011). Welfare Quality® researchers also concluded that composite labels including all these dimensions may suit consumers' concerns more than a label that is dedicated solely to animal welfare (Kjaernes and Larvik, 2008). An example can be found in the French system 'Label Rouge' which primarily aims to guarantee a higher product quality but which has also begun to include environmental and animal welfare issues (see http://www.volaillelabelrouge.com/012_volaille_LR.php). Clearly, it could be extremely difficult to attain excellent scores on all aspects (i.e. welfare, environment, product quality).

On the other hand, some production/certification schemes, such as Freedom Food, are focused solely on animal welfare (http://www.rspca.org.uk/freedomfood/aboutus). In such schemes one could expect that the highest welfare standards are met, although this should always be checked. The Welfare Quality® 'excellent' category seems particularly relevant for such schemes.

9.5.3 Monitoring progress

A welfare assessment could also be used by an external advisor to monitor the results of welfare improvement strategies applied by the farmer and to provide him/her with the appropriate feedback. Although the full range of results from the Welfare Quality® assessment system could be useful it is likely that criterion scores or even the results of one or two specific measures would be more relevant to focus on specific problems identified in an earlier assessment (see also the example on bird cleanliness in Frame 1). It is also conceivable that the Welfare Quality® assessment system could be used as a 'self-assessment' management tool by the farmer or processor to help identify welfare problems or risk factors and to monitor the effects of any improvement strategies applied.

9.5.4 Assessing new animal farming systems/breeds

The Welfare Quality® assessment system can also be used to evaluate the impact of new farming systems and/or different husbandry practices or devices on the welfare status of the animals. For instance, a systematic evaluation of new husbandry devices is already routinely carried out in Switzerland and only those devices that meet certain requirements are allowed (Wechsler, 2001; 2003). Additionally, the Welfare Quality® system could be used to check the existing welfare status of emerging breeds. Regardless of whether such assessments form part of commercial R&D or of specific research programmes (Bracke, 2001; Bracke *et al.*, 2007; Veissier *et al.*, 2004), it would be far preferable to apply the same standardised methods such as those of Welfare Quality®, so that meaningful comparisons can be made.

9.5.5 Increased transparency of welfare information

An interactive web platform was developed to provide users with various levels of information on the Welfare Quality® measures, the calculation of scores and the overall welfare assessment (http://www.clermont.inra.fr/wq/). For example, users can now see the distribution of the results of welfare assessments according to the various animal types and the year and country in which the assessments were carried out. Farmers are also given access to their own data so that they can compare their results to those obtained at other farms in the same population (benchmarking); this enables them to determine how well or badly they are performing in relation to other farmers. Simulation of results is also possible: producers can take their own data and simulate either an increase or a decrease in one or several of the welfare measures; they can then see the consequences of the simulated changes on their criterion and principle scores and on the overall assessment. This exercise can help to guide farmers' decisions regarding the likely effectiveness of particular welfare improvement strategies, such as alterations to environmental features or management practices. Because different types of user (e.g. producers, assessors, certification bodies, retailers, researchers, NGOs, etc.) can access the platform software several user-profiles were defined, each with specific access rights.

9.5.6 Provision of targeted support and advice

To illustrate this targeted application of the Welfare Quality® system we focus on a specific example, in this case how to help a dairy farmer resolve a problem of lameness in his cows. A structured assessment and analysis will let him know how his/her performance compares to that of other farms of a similar type; this type of benchmarking exercise will also help the farmer to identify likely causes and/or risk factors for the lameness problem. In the next (advisory) stage the farmer can be shown how and where to access important resources (background information, remedial treatments, farming support agencies, etc.) that can help him/her to tackle the lameness problem. For instance, specific information gained during the Welfare Quality® assessment on the type of flooring at the farm and on the hoof care strategy used could indicate the most appropriate remedial solution, e.g. change the flooring, monitor hoof health more frequently, etc. Because lameness not only reduces the cows' welfare but can also have damaging economic effects in terms of reduced productivity, increased veterinary bills, etc., using the Welfare Quality® information in the most effective way is likely to result in measurable economic benefits. In other words, targeted improvement strategies can help both the farmer and the animal. Of course, to be viable, remedial strategies must satisfy not only welfare and economic requirements but they would also have to be practicable, i.e. safe, affordable and easy to implement by the farmer and/or breeding company.

The Welfare Quality® measures could also have substantial value in providing Key Welfare Indicator information to veterinarians who can then include this information in their interaction with their farmer clients.

Furthermore, in all countries of the EU, the state, usually through a state veterinary service carries out some visits to farms to ensure that they can deliver compliance with animal care and welfare legislation. It is possible that a previous demonstration of adequate (or even high) welfare standards would allow subsequent farm inspections to be targeted on specific issues. In this respect the state inspection work load could be optimised by using information provided by harmonised assessment schemes such as Welfare Quality®. The potential for streamlining of animal health and welfare visits/inspections could be of real benefit to farmers and to the state.

One point we must always remember is that the relationship between the farmer and the assessor is likely to be critical to the application of animal welfare assessments. The assessor should not be viewed as a policeman but rather as someone who can offer genuine help.

References

Bracke, M.B.M. (2001). Modelling of animal welfare. The development of a decision support system to assess the welfare status of pregnant sows, Wageningen University, Wageningen, the Netherlands, 150 pp.

Bracke, M.B.M., Zonderland, J.J. and Bleumer, E.J.B. (2007). Expert consultation on weighting factors of criteria for assessing environmental enrichment materials for pigs. Applied Animal Behaviour Science, 104, 14-23.

European Commission (1998). Council Directive 98/58/EC of 20 July 1998 concerning the protection of animals kept for farming purposes. Official Journal of the European Union L 221, 8.8.1998, 23-27.

European Commission (1999). Council Directive 1999/74/EC of 19 July 1999 laying down minimum standards for the protection of laying hens. Official Journal of the European Union L 203, 3.8.1999, 53-57.

European Commission (2001a). Council Directive 2001/88/EC of 23 October 2001 amending Directive 91/630/EEC laying down minimum standards for the protection of pigs. Official Journal of the European Union L 316, 1.12.2001, 1-4.

European Commission (2001b). Commission Directive 2001/93/EC of 9 November 2001 amending Directive 91/630/EEC laying down minimum standards for the protection of pigs. Official Journal of the European Union L 316, 1.12.2001, 36-38.

European Commission (2007). Council Directive 2007/43/EC of 28 June 2007 laying down minimum rules for the protection of chickens kept for meat production. Official Journal of the European Union L 182, 12.7.2007, 19-28.

European Commission (2008). Council Directive 2008/119/EC of 18 December 2008 laying down minimum standards for the protection of calves. Official Journal of the European Union L 10, 15.1.2009, 7-13.

European Commission (2009a). Impact assessment report. Commission staff working document accompanying the report from the Commission to the European Parliament, the Council, the European Economic and Social Committee and the Committee of the Regions Options for animal welfare labelling and the establishment of a European Network of Reference Centres for the protection and welfare of animals, SEC(2009) 1432, pp. 1-86.

European Commission (2009b). Council Regulation (EC) No 1099/2009 of 24 September 2009 on the protection of animals at the time of killing. Official Journal of the European Union L 303, 18.11.2009, 1-30.

Kjærnes, U. and Lavik, R. (2008) Opinions on animal welfare and food consumption in seven European countries. In: Kjærnes, U., Bock, B.B., Roe, E. and Roex, J. (eds.) Consumption, distribution and production of farm animal welfare. Welfare Quality Reports no.7, Cardiff University, Cardiff, UK, pp. 3-128.

Miele, M., Veissier, I. and Evans, A. (2011). Animal welfare: establishing a dialogue between science and society. Animal Welfare, 20, 103-117.

Polten, R. (ed.) (2007). Proceedings of the workshop 'Animal welfare improving by labelling?', 28 March 2007, Brussels, Belgium, 86 pp.

Veissier, I., Capdeville, J. and Delval, E. (2004). Cubicle housing systems for cattle: comfort of dairy cows depends on cubicle adjustment. Journal of Animal Science, 82, 3321-3337.

Wechsler, B. (2001). Pretesting of mass-produced farm animal housing systems in Switzerland 20 years of experience. In: Kuczyński, T. and Sällvik, K. (eds.) Proceedings of the nternational Symposium of the 2nd Technical Section of C.I.G.R. on Animal Welfare Considerations in Livestock Housing Systems, Szklarska Poreba, Poland, pp. 55-67.

Wechsler, B. (2003). Testing of mass-produced farm animal housing systems with regard to animal welfare. In: Van der Honing, Y. (ed.) Proceedings of the 54th Annual Meeting of the European Association for Animal Production, Wageningen Academic Publishers, Wageningen, the Netherlands, p. 125.

Chapter 10. Assessing and improving farm animal welfare: the way forward

Harry Blokhuis, Bryan Jones, Mara Miele and Isabelle Veissier

10.1 Introduction

The present chapter describes several approaches (as well as some essential requirements and conditions) to facilitate and support a harmonised and effective implementation of the Welfare Quality® assessment systems for diverse purposes. Even though Welfare Quality® was the largest ever collaborative project in animal welfare science, it could not possibly have covered all the questions and every detail (Blokhuis *et al.*, 2010); indeed it generated new questions (Miele *et al.*, 2011). So, it is not surprising that there are still unanswered questions and discussion points. These relate for instance to the development of animal-based welfare measures where these are currently not available, the necessary frequency of assessments, how to integrate new knowledge, and how to tackle practical issues of implementation. One particular obstacle to the widespread application of the existing Welfare Quality® protocols is the relatively large amount of time and effort needed for a complete welfare assessment on farm. Several follow-up initiatives now focus on exploring the feasibility of different ways of simplifying the protocols or finding shorter but efficient ways to implement the system. Firstly, for example the identification of reliable and meaningful 'sentinel' indicators, i.e. indicators that are likely to reveal major problems on a farm, or the design of risk models based on information on the living conditions of animals, could lead to more efficient assessments and/or the use of risk based audits with a farm visit oriented primarily or even exclusively towards major risks. Another line of current research investigates various methods of reducing the workload and time required by replacing some of the manual measures with automated ones, perhaps by using modern technology like sensors, sensing systems (image, sound, etc.) and real time modelling (Berckmans, 2008).

The implementation of an assessment system occurs in an environment that is strongly influenced by economic, political, technological and socio-cultural factors which can all interact with each other. This, together with the various possible approaches, requirements and conditions as well as the on-going research and development creates enormous complexity and causes some uncertainty among stakeholders (c.f. Ingenbleek *et al.*, 2011). A trustworthy governance institution that guides the implementation process and guarantees harmonisation, continuity and transparency is crucial to provide actors in animal production chains with the necessary confidence to invest in the implementation and use of the Welfare Quality® assessment systems.

The main areas of concern for an effective implementation of the Welfare Quality protocols as identified above as well as potential solutions are briefly addressed below. The chapter closes with a brief paragraph summarising the need for a holistic approach to safeguarding and progressing animal welfare.

10.2 Governance of development and implementation

Scenario analyses carried out within Welfare Quality® (Ingenbleek *et al.*, 2011) also emphasised the importance of establishing an independent body or an institution to facilitate the implementation of a harmonised animal welfare assessment system. Such an institution would have strategic responsibilities for developing a common vision on how to support and manage such implementation for the various species. Moreover, in the context of other contemporary contested issues (e.g. sustainability) the need for new kinds of institutions to coordinate policy and guide innovation and development in industry was highlighted (Lundvall *et al.*, 2002).

The idea of a *European Network of Reference Centres for Animal Protection and Welfare* as proposed by the European Commission (2009) could possibly play an important role in the governance and harmonisation of a European assessment system. The EC's communication on the European Union strategy for the protection and welfare of animals 2012-2015 (European Commission, 2012) mentions that the provision of coherent and uniform technical information in the context of outcome-based animal welfare indicators would be an important task of such a network.

Activities to be addressed in this context are:
- managing the assessment systems and support instruments;
- maintaining and upgrading the assessment systems and the support tools;
- ensuring support and acceptance among stakeholders;
- prioritising and facilitating research.

A coordinated European animal welfare network would be admirably suited to fulfil all of the above roles. Given that there is considerable variation both in environment and production practices within Europe and that specific expertise is available in different Member States, such a network could clearly create valuable synergies and could benefit from the consideration and/or incorporation of existing national practices and information.

The outcomes of the recently started EUWelNet project are considered likely to inform decisions on the establishment of such a European coordinating institution.

One aim of the project is 'to establish proof of principle for a coordinated European animal welfare network' and another 'to conclude on the feasibility and the possible conditions under which the Union could support a European coordinated network for animal welfare'

10.2.1 Management of the system and support instruments

To facilitate a harmonised and effective implementation of the welfare assessment systems and to avoid confusion and conflicts in the market, it is essential that an authoritative institution defines the conditions of use and describes the sorts of claims that can be made on the basis of the outcomes of the assessments.

Since the assessor is a 'critical component' of every certification and inspection scheme, appropriate and recognised training in the use and practical application of the welfare assessment protocols is essential to ensure uniform scoring. Moreover, assessors should be regularly re-evaluated when they are active in the field to ensure retention of objectivity, impartiality and repeatability in scoring (Gibbons *et al.*, 2012). Quality control and harmonised pass/fail criteria are essential elements that should be provided by a governing institution.

The application (by trained assessors) of Welfare Quality® assessment systems in livestock production will generate data from individual farms all over Europe (and beyond). The control of such data and the provision of advice regarding the correct handling of the data and its processing into integrated welfare assessment scores is an important quality aspect. The large amounts of data (results from measures and calculated scores) from the above activities should preferably be centrally stored to ensure the safe and steady accumulation of knowledge. Data collected at several locations and at various intervals can subsequently be used (with appropriate protection of private and commercial interests) to: (1) continue to inform stakeholders (e.g. on the progress made by a certain population of farms / certain slaughterhouses, etc.); (2) help farmers or slaughterhouse managers to see the progress they are making and to compare their status with industry averages (i.e. benchmarking); and (3) further analyse the links between welfare problems and identify their associated risk factors. The database could also be used for a yearly 'European welfare barometer' with statistical summaries of assessment scores according to species, housing system and management.

10.2.2 Maintenance and upgrading of the system

The Welfare Quality® protocols were published at the end of the project (December 2009). These received a lot of attention and were generally very well received, indeed they are now being used in numerous scientific studies of farm animal welfare. Furthermore, members of the Welfare Quality Network are actively examining ways of upgrading the protocols (see www.welfarequalitynetwork.net and Chapter 9). To facilitate the implementation of the assessment protocols in different EU Member States, translation into other major languages would be very supportive. One role of a governing institution would be to ensure the correctness, quality and effective dissemination of such translations.

The existing Welfare Quality® assessment protocols and evaluation models need to be regularly updated and refined on the basis of new scientific findings, societal developments, and practical experiences gained during implementation. For example, some new measures may prove easier to collect, or they may be more precise or more reliable than some contained in the current Welfare Quality® protocols. A governing structure should coordinate and check the stringent testing of the validity, repeatability and robustness of a new measure as well as the process required to translate the data into an integrated value score.

Currently, the Welfare Quality® scoring system proposes that animal units should be placed in one of four welfare categories according to the following rules:
- 'Excellent', requires a score of at least 55[16] on all 4 welfare principles and 80 on 2 of them;
- 'Enhanced', requires at least 20 on all welfare principles and 55 on 2 of them;
- 'Acceptable', requires at least 10 on all welfare principles and 20 on 3 of them;
- 'Not classified' is failure to meet the above requirements.

The results of the citizen juries that were carried out in three key EU countries (United Kingdom, Norway and Italy) pointed out the higher expectations of the EU public in terms of the level of animal welfare achieved in each of the above categories (Evans and Miele, 2012; Miele *et al.*, 2011). On the basis of these results it was therefore recommended that after some years of implementation and the achievement of visible improvements at farms and abattoirs, the categorisation requirements should be increased to:
- 'Excellent', at least 55 on all principles and 80 or more on 2 of them;
- 'Enhanced', at least 20 on all principles and 55 or more on 3 of them;
- 'Acceptable', minimum of 20 on all principles;
- 'Not classified' is failure to meet the above requirements.

Based on the collated assessment data a governing institute should establish what welfare levels are actually achieved for the different measures on farms and in

---

[16] On a 0-100 value scale.

slaughterhouses. Such data represent essential input to a consultation and negotiation process with stakeholders to determine whether there has been sufficient welfare improvement to allow implementation of the second set of rules.

10.2.3 Helping to ensure solid acceptance among stakeholders

The active participation of a broad range of stakeholders (farmers, breeders, retailers, certification bodies, NGOs, national governments, etc.) in the actual research and in an advisory capacity as well as the extensive investigation of consumers' knowledge and practices greatly facilitated the acceptance and uptake of Welfare Quality® outcomes. Stakeholder participation and involvement should also be secured during the further development of practical welfare indicators and feasible assessment systems. An example of a type of structure that could facilitate such involvement is the European Animal Welfare Platform (EAWP). This platform originated from the Welfare Quality® project and was an unique and innovative stakeholder partnership comprising key industry players in the food chain, animal welfare organisations and welfare science. Working together in an atmosphere of openness and trust the members of the platform facilitated the exchange of knowledge, expertise, resources and networks to identify and prioritise key welfare issues/problems in several animal product groups (beef and dairy cattle, pigs, laying hens, broiler chickens, and salmon) and how to assess these. The EAWP then described: the consequences of each of the identified welfare problems, ways in which they could be monitored and measured, existing best practices for dealing with each welfare issue, and proposed short- and long-term goals for welfare and economic improvement as well as a list of the most pressing R&D priorities. These outcomes are gathered in a set of Strategic Approach Documents for each of the product groups (available on www.animalwelfareplatform. eu). Platforms such as the EAWP could clearly play an important role in involving stakeholders in the governing process and to assure broad acceptance of assessment systems.

The provision of sound advice on ways of avoiding welfare risks or resolving problems is critical for the uptake and implementation of the Welfare Quality® assessment systems by end users and for improving farm animal welfare in general. Potential remedial measures developed within and outside the Welfare Quality® project are described in a Technical Information Resource (Jones and Manteca, 2009; www. welfarequalitynetwork.net) which also details the causes and consequences of welfare problems. Also here, a governance structure is necessary to maintain, update (as new results emerge), extend (to include new strategies and new species) and disseminate this valuable resource.

10.2.4 Prioritising and coordinating research

Through experience gained during the management of the assessment systems and the efforts to support implementation, areas requiring further research are expected to be highlighted. Such areas may include: fundamental biological knowledge required to assess, safeguard and improve all aspects of animal welfare as well as relevant developments in public concerns about animal welfare, stakeholder or consumer attitudes, areas of cross-compliance, and social, economic and environmental policy decisions.

A broad-based prioritisation of such research and development needs would help a range of European and national research funding bodies to focus effectively on the most relevant issues and support informed decision-making.

10.3 Concluding remarks

We believe that the delivery of the Welfare Quality® assessment and information systems as well as the welfare improvement strategies can be considered valuable and innovative developments. The many achievements of Welfare Quality® would not have been obtained without the intensive interaction that took place between animal scientists, social scientists and stakeholders. To the best of our knowledge there had been no previous initiative that combined expertise from the social and natural sciences, ethics and mathematics as well the participation of a broad range of stakeholders to design a practical tool for the assessment of animal welfare. This high level of interaction was not always self-evident at the start, probably because people came from different backgrounds and had different research habits, different interests and different priorities. However, the various tools (e.g. working groups, advisory boards, etc.) that were used in Welfare Quality® to stimulate and organise these interactions were successful and resulted in a truly cooperative and fruitful way of working. Indeed, thanks to these many interactions the outcomes can be said to be based on a wide consensus, to be exhaustive and at the same time to formalise value judgements.

We can already see various consequences of the Welfare Quality® project and its outcomes as they relate to research, assurance schemes and the animal industry. Firstly for example, in animal welfare science many researchers take Welfare Quality® as a standard and develop similar systems in other species using the Welfare Quality® approach (e.g. horses, sheep and goats). Secondly, Welfare Quality® outcomes are being taken on board and animal-based parameters are increasingly included in further developments of insurance schemes (e.g. AssureWel, www.assurewel.org). Thirdly, various industry partners are already exploring the implementation of Welfare Quality®

systems. An inventory in 2011 of work carried out by the partners of the Welfare Quality Network (see below) revealed that there were more than 40 projects (either running or planned) that were based on the findings of the Welfare Quality® project. About 20 of these projects involved close collaboration with industrial partners.

The importance of retaining the partnership and expertise established in the Welfare Quality® project has already been highlighted in numerous official communications (e.g. Report from the EU Commission (IP/09/1610); Report of the Committee on Agriculture and Rural Development (2009/2202(INI)); speeches by the EU Commissioner for Health and consumer protection and representatives of the EU Directorate General for Research during the final conference of the Welfare Quality® project in October 2009). More than 25 of the original partners of Welfare Quality® partners therefore took the initiative to form a collaborative structure (the Welfare Quality Network, WQN (see above and www.welfarequalitynetwork.net). Its main objectives are to ensure harmonisation of high quality animal welfare assessment measures and systems by maintaining and upgrading the Welfare Quality® assessment protocols and supporting stakeholders in their implementation trajectories and practical procedures such as data capture, the calculation of overall welfare scores, training, etc.

It is particularly gratifying to look back on the achievements of the Welfare Quality® project and to see that they will not simply sit on a shelf and gather dust. The way forward is clear and several avenues are already being explored.

References

Berckmans, D. (ed.) (2008). Precision livestock farming (PLF). Computers and Electronics in Agriculture, 62, 1-80.

Blokhuis, H.J., Veissier, I, Miele, M. and Jones, R. B. (2010). The Welfare Quality® project and beyond: safeguarding farm animal well-being. Acta Agriculturae Scandinavica A, Animal Science, 60, 129-140.

European Commission (2009). Report from the Commission to the European Parliament, the Council, the European Economic and Social Committee and the Committee of the regions: Options for animal welfare labelling and the establishment of a European Network of Reference Centres for the protection and welfare of animals. COM (2009) 584 final.

European Commission (2012). Communication from the Commission to the European Parliament, the Council and the European Economic and Social Committee on the European union strategy for the protection and welfare of animals 2012-2015.

Evans, A. and Miele, M. (2012). Between food and flesh: how animals are made to matter (and not to matter) within food consumption practices, Environment and Planning D – Society and Space, 30(2), 298-314.

Ingenbleek, P.T.M., Blokhuis, H.J., Butterworth, A. and Keeling, L.J. (2011). A scenario analysis on the implementation of a farm animal welfare assessment system. Animal Welfare 20, 613-621.

Gibbons, J., Vasseur, E., Rushen, J. and de Passillé, A.M. (2012). A training program to ensure high repeatability of injury scoring of dairy cows. Animal Welfare, 21: 379-388.

Jones B. and Manteca, X. (2009). Developing practical welfare improvement strategies. In: Keeling L.M. (ed.) An overview of the development of the Welfare Quality® project assessment systems. Welfare Quality Reports No. 12, Cardiff University, Cardiff, UK, pp. 57-65.

Lundvall, B.A., Johnson, B., Andersen, E.S. and Dalum, B. (2002). National systems of production, innovation and competence building. Research Policy 31, 213-231.

Miele, M., Veissier, I., Evans, A. and Botreau, R. (2011). Animal welfare: establishing a dialogue between science and society, Animal Welfare, 20, 103-117.

Appendix 1. Partners in the Welfare Quality® project

Participant name	Country	Website
University of Natural Resources and Applied Life Sciences Vienna	Austria	http://www.boku.ac.at
University of Veterinary Medicine	Austria	www.vu-wien.ac.at
Katholieke Universiteit Leuven	Belgium	www.kuleuven.be
Instituut voor Landbouw- en Visserijonderzoek	Belgium	http://www.ilvo.vlaanderen.be/
Department of Animal Science, Faculty of Agriculture and Veterinary Sciences, Brazil	Brazil	www.unesp.br
Veterinary Faculty, Universidad de Chile	Chile	http://www.uchile.cl/
Institute of Animal Science	Czech Republic	www.vuzv.cz
Aarhus University	Denmark	http://www.au.dk/en/
University of Copenhagen	Denmark	http://www.ku.dk
Institut du Porc	France	http://www.ifip.asso.fr
Coopérative Interdépartementale Aube, Loiret, Yvonne, Nièvre	France	www.cialyn.fr
Institut National de la Recherche Agronomique	France	http://www.inra.fr
Institut de l'Elevage	France	http://www.inst-elevage.asso.fr
Institut Supérieur d'Agriculture Lille	France	www.isa-lille.fr
UPRA France Limousine Selection	France	www.limousine.org
Université Pierre et Marie Curie (Paris 6 University)	France	http://www.upmc.fr
University of Toulouse II – le Mirail	France	http://www.univ-tlse2.fr/
University of Kassel	Germany	http://www.uni-kassel.de
Teagasc – The National Food Centre	Ireland	http://www.teagasc.ie
Università degli Studi di Milano	Italy	www.unimi.it
Università degli Studi di Parmai	Italy	www.unipr.it
Università degli Studi di Padova-Dipartimento di Scienze Animali	Italy	http://www.dsa.unipd.it/
Università di Pisa	Italy	www.unipi.it
Centro Ricerche Produzioni Animali SpA	Italy	www.crpa.it
Faculty of Veterinary Medicine, Mexico	Mexico	http://www.fmvz.unam.mx/
Wageningen UR Livestock Research	Netherlands	www.Wageningenur.nl/livestockresearch
Wageningen University	Netherlands	http://www.wageningenur.nl

National Institute for Consumer Research	Norway	http://www.sifo.no/
Norwegian Agricultural Economics Research Institute	Norway	www.nilf.no
Norwegian University of Life Sciences	Norway	http://www.umb.no/
Institut de Recerca i Tecnolgia Agroalimentàries	Spain	www.irta.es
Universitat Autònoma de Barcelona	Spain	www.uab.es
Sveriges Lantbruksuniversitet	Sweden	http://www.slu.se
Goteborg University	Sweden	http://www.gu.se
Lund University	Sweden	www.fek.lu.se
Stockholm University	Sweden	http://www.statsvet.su.se
Cardiff University	United Kingdom	http://www.cardiff.ac.uk/
Scotland's Rural College	United Kingdom	http://www.sruc.ac.uk
Newcastle University	United Kingdom	http://www.ncl.ac.uk
University of Bristol	United Kingdom	http://www.bristol.ac.uk
University of Reading	United Kingdom	http://www.reading.ac.uk/
University of Exeter	United Kingdom	http://www.exeter.ac.uk/
Universidad de la República Uruguay	Uruguay	http://www.universidad.edu.uy/

List of authors

R. M. Bennett, School of Agriculture, Policy and Development, University of Reading, Whiteknights, P.O. Box 237, Reading RG6 6AR, United Kingdom

H.J. Blokhuis, Swedish University of Agricultural Sciences, Department of Animal Environment and Health, Box 7068, 750 07 Uppsala, Sweden

B. Bock, Wageningen University, Rural Sociology, Hollandseweg 1, 6706 KN Wageningen, the Netherlands

R. Botreau, INRA, UMR1213 Herbivores, 63122 Saint-Genes-Champanelle, France; and Clermont Université, VetAgro Sup, UMR1213 Herbivores, BP 10448, 63000 Clermont-Ferrand, France

H. Buller, University of Exeter, Department of Geography, Amory Building, Rennes Drive, Exeter EX4 4RJ, United Kingdom

A.Butterworth, Bristol University, Division of Farm Animal Science, Clinical Vet Science, Langford, N Somerset, BS40 5DU, United Kingdom

A.Dalmau, IRTA, Finca Camps i Armet s/n, 17121 Monells (Girona), Spain

A. Evans, University of Essex, Department of Sociology, Wivenhoe Park, Colchester CO4 3SQ, United Kingdom

B. Forkman, University of Copenhagen, Division of Ethology, Department of Large Animal Sciences, Faculty of Health and Medical Sciences, Grønnegårdsvej 8, 1870 Frederiksberg, Denmark

R.B. Jones, Animal Behaviour & Welfare Consultancy, 110 Blackford Ave, Edinburgh EH9 3HH, United Kingdom

L. J. Keeling, Swedish University of Agricultural Sciences, Department of Animal Environment and Health, P.O. Box 7068, 750 07 Uppsala, Sweden

U. Kjaernes, National Institute for Consumer Research (SIFO), P.O. Box 4682, Nydalen, 0405 Oslo, Norway

X. Manteca, Universitat Autonoma de Barcelona, School of Veterinary Science, Campus de la UAB, 08193 Bellaterra Barcelona, Spain

M. Miele, Cardiff University, School of Planning and Geography, Glamorgan Building, King Edward VII AVE, Cardiff CF 103 WA, United Kingdom

I. Veissier, INRA, UMR1213 Herbivores, F-63-122 Saint-Genes-Champanelle, France; and Clermont Université, VetAgro Sup, UMR1213 Herbivores, BP 10448, 63000 Clermont-Ferrand, France

A. Velarde, IRTA, Finca Camps i Armet s/n, 17121 Monells (Girona), Spain

C. Winckler, University of Natural Resources and Applied Life Sciences – BOKU, Department of Livestock Sciences, Animal Husbandry Group, Gregor-Mendel-Straße 33, Vienna 1180, Austria

Index

Improving farm animal welfare